Nano Hybrids and Composites
Volume 33

Scientific.Net

Nano Hybrids and Composites
ISSN: 2297-3370, Vol. 33, pp 1-11
© 2021 Trans Tech Publications Ltd, Switzerland

Submitted: 2021-04-05
Revised: 2021-07-26
Accepted: 2021-07-29
Online: 2021-10-11

Fabrication and Characterization of Composite Materials Using Multiple Waste Materials (Leather & Jute Fabrics) and Unsaturated Polyester Resin

Md. Farhad Ali[1,2,4,a*], Md. Sahadat Hossain[2,b], Samina Ahmed[2,3,c] and A.M. Sarwaruddin Chowdhury[4,d]

[1]Institute of Leather Engineering and Technology, University of Dhaka, Dhaka-1000, Bangladesh

[2]Glass Research Division, Institute of Glass & Ceramic Research and Testing, Bangladesh Council of Scientific and Industrial Research (BCSIR), Dhaka-1205, Bangladesh

[3]BCSIR Laboratories Dhaka, Bangladesh Council of Scientific and Industrial Research (BCSIR), Dhaka-1205, Bangladesh

[4]Department of Applied Chemistry and Chemical Engineering, Faculty of Engineering and Technology, University of Dhaka, Dhaka-1000, Bangladesh

[a]farhad.ilet@du.ac.bd, [b]saz8455@gmail.com, [c]shanta_samina@yahoo.com, [d]sarwar@du.ac.bd

*Corresponding author: Md. Farhad Ali; E-mail: farhad.ilet@du.ac.bd

Keywords: Leather waste; waste jute fabrics; composite; mechanical properties; resin

Abstract. Nowadays, the environment is getting polluted due to different types of manmade pollutants because of the excess use of synthetic materials. Various kinds of waste materials, which are eluted from numerous industries, are also enhancing environmental pollution. So, utilization of waste materials along with reduction of using synthetic materials will definitely subside the environmental pollution. In this research, waste jute fabric and waste leather (cow hides) were used as reinforcing agents and unsaturated polyester resin (UPR) as a matrix to fabricate environmentally friendly composite materials. Hand-lay up technique was conducted to prepare composite materials. Different percentages of waste leather and used jute fabrics were used with the UPR to modify properties. Improved mechanical properties such as tensile strength (TS), tensile modulus (TM), and percentage elongation at break (EB) have been experimented for every percentage of reinforcement addition, and 2% leather along with 10% jute fabrics containing sample revealed the best result (TS=24MPa, TM=907MPa, and EB=0.98%). Composites were also characterized by the scanning electron microscope (SEM) and Fourier transform infrared (FT-IR) to support the results in favor of enhanced mechanical properties.

Introduction

Now, researchers are more concerned regarding environmental pollution due to the change in climate and natural calamity. Sustainable development, environment, and economy are the most taking issue throughout the world. To make the environment pollution-free, it is needed to reduce the use of synthetic polymer as well as to reuse the waste materials. Leather industries are rapidly growing industries in the Asia Pacific region and thus these industries are also adding waste materials to the environment causing pollution [1-3]. If these waste materials cannot be processed into commercial materials, they will be problematic for the industry. Shaving waste is also a waste material that is obtained from the cowhides and is responsible for environmental pollution. In this Asia Pacific region natural fiber, mainly jute fiber is cultivated to a great extend. Bangladesh and India are the two most jute fiber-producing countries. Thus, the products from the jute are available and comparatively cheap to other synthetic products. After finishing the lifetime of these jute-based products, they are directly thrown out into the environment. Though the jute fibers are completely biodegradable, they can pollute the environment by microbial decomposition especially when large quantities are store in a particular place. Sometimes water bodies can be polluted by higher nutrients from the degradation of biodegradable jute fabrics. Thus the reuse of jute products can be a good alternative to minimize environmental pollution. Thus, the recycling of jute products is more economical than disposal [4]. The constituting compounds of jute fibers are cellulose (61–71%), hemicellulose (13.6–20.4%), lignin (12–13%), ash (0.5–2%), pectin (0.2%), wax (0.5%), and moisture (12.6%) [5–8]. Among these compounds, cellulose is mainly responsible for the strength of jute fibers. These high cellulose-based jute fibers are used in composite fabrication to give strength as the reinforcing material [9-12]. When we are talking about the waste jute fiber, all the compounds except cellulose are degraded or washed out. Celluloses are responsible for incorporating strength in composites. Composites materials are drawing attention to the researcher due to their modifiable properties and versatile uses. Fiber-reinforced composites can be used in different applications like insulation for electrical construction, boxes, covers, cable tracks, gears, bearings, machine rods, pipes, machine rods, pipes, chimneys, housing cells, furniture, bathrooms, robot arms, protection helmets, etc [8, 13,14]. In all these applications the synthetic fiber can be replaced by natural fiber. The use of natural fiber in these sectors will minimize environmental pollution. Jute fiber-reinforced composites are widely used in the mentioned sectors. But, waste jute and leather-based composites are new in the field of materials science. Most of the documented fiber-reinforced composites are related to the waste jute fiber-reinforced composites. Waste jute fibers are used as a reinforcing agent in composites fabrication. The addition of certain percentages of jute is improving the physical, mechanical, thermal, etc. properties of composite materials [15-16]. Hybrid composites are also drawing attention to the researcher, which is used to modify the properties and to add mechanical strength as well as lower the production cost [17-19].

In this research, we used waste jute products as reinforcing agents and waste leather as additives. Both of these materials are directly related to the clean environment. UPR was used as the matrix material. The addition of waste jute and waste leather will reduce environmental pollution and curtail the production cost. The use of leather waste and waste jute products is the unique feather in this research.

Materials and Methods

Raw Materials

Shaving waste was collected from the local tannery industries near Hemayetpur, Dhaka, Bangladesh. A considerable amount of leather wastes are produced during every operation of transforming raw hides and skins to leather as a final product. Each step leads to the generation of different kinds of wastes. During each stage of processing, the tannery generates a considerable amount of solid wastes. For this study, we have considered only shaving wastes from wet blue tanned leather which is portentous, impregnated with chromium, synthetic fat, oil, tanning agents, and these shaving wastes were collected from the leather industry, hemayetpur. These leather wastes are ground by a graining machine, which is illustrated in Figure 1. Jute, commonly referred to as the 'golden fiber' is a commercial crop grown for its fiber. The fibers are extracted from the plant stem which usually grows to a length of 1-4 meters. Jute fiber has a density of 1300 Kg/m^3 and tensile strength of 583 MPa. The Young's modulus of these fibers is 26.5 GPa. Jute fibers are extensively used in the textile industry and for the production of mats and bags due to their good strength and tenacity. Recycled jute stacks were purchased from a grocery store as shown in Figure 2. The jute sacks had previously been used to store peanuts, dried chilies, or various beans. The quality of the jute sacks varied as there were some contaminants (woodchips, string, dust) embedded in the sacks that were damaged (holes, broken fibers) in some areas. The jute stacks were dirty and, sometimes, visibly damaged. All the jute sacks were machine washed and sun-dried before use. Physical treatment was done and the fiber was chopped manually to desired dimensions. Collected jute fibers were cut by hand, and the cut fibers were washed and dried. The jute reinforcements used in the project are collected from waste jute sacks. Unsaturated polyester resin (UPR) and Methyl ethyl ketone peroxide (MEKP) were obtained from LUCKY ACRYLIC LTD with properties such as appearance (pink), density (1.12 g/cm3), and stability in the dark below 25°C.

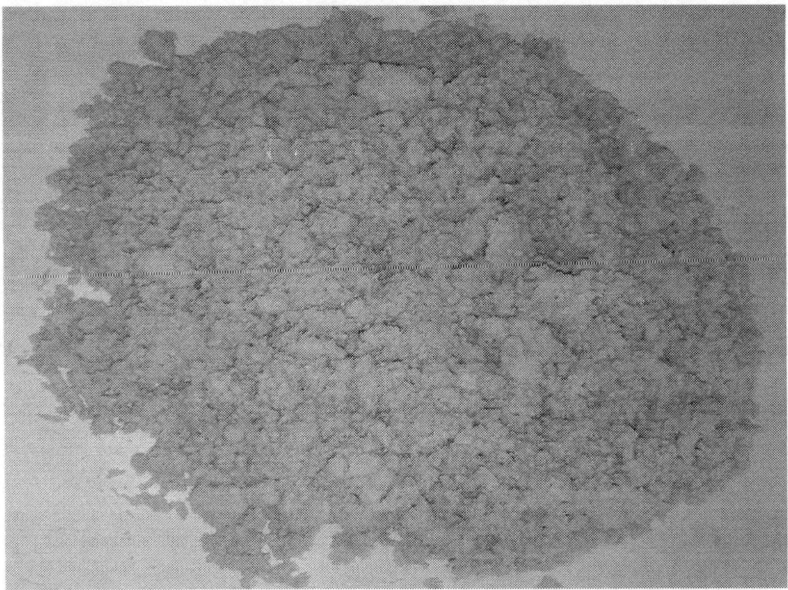

Figure 1: Grinded shaving waste

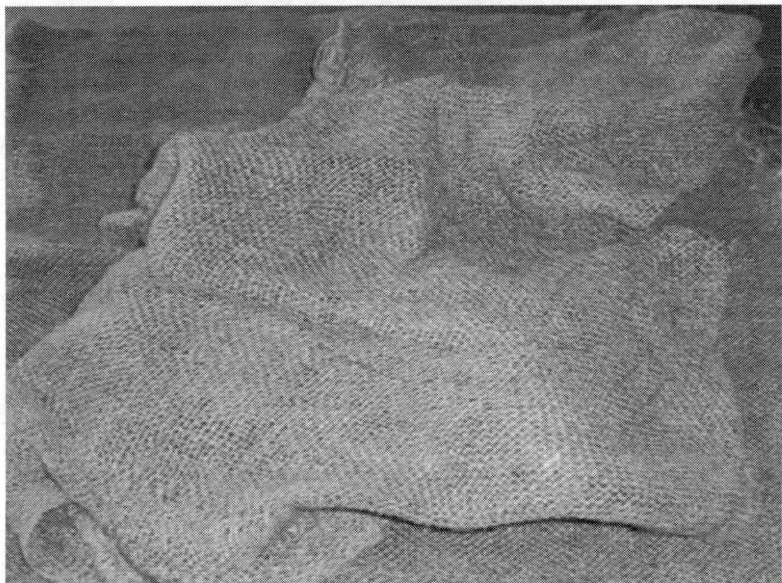

Figure 2 : Recycled jute sacks

Composite fabrication

The composite fiber was prepared by hand lay-up technique. A mold with dimensions of 20 x 10 × 2 cm was prepared for composite fabrication. For different volume fractions of fibers, a calculated amount of UPR was taken in a glass beaker, and the composition is depicted in table 1. Shaving waste and jute fibers were mixed with UPR into the beaker. After mixing for 15 minutes 1% hardener was added to the mixture in the beaker. After keeping the mold on a ply board the mixture was poured. Care was taken to avoid the formation of air bubbles. The pressure was then applied from the top, and the mold was allowed to preserve at room temperature for 72 hrs. During the application of pressure, some amount of mixture of UPR and hardener squeezes out. Care has been taken to consider this loss during the manufacturing of composite sheets. After 72 hrs, the samples were taken out from the mold, cut into different sizes, and kept in an air-tight polyethylene bag for further testing.

Table 1: Composition of different types of samples

Sample No.	Leather dust	Jute	UPR	MEKP
1	2%	2.5%	94.5%	1%
2	2%	5%	92%	1%
3	2%	7.5%	89.5%	1%
4	2%	10%	87%	1%
5	2%	12.5%	84.5%	1%
6	2%	15%	82%	1%
7	2%	17. 5%	79.5%	1%
8	2%	20%	77%	1%
9	2%	22.5%	74.5%	1%
10	2%	25%	72%	1%

Scanning Electron Microscopy (SEM)

The surface morphologies of composites were examined by scanning electron microscopy (SEM) model Phonom Pro. Samples were mounted with carbon tape on aluminum stubs. The composite specimens were examined to observe the morphology of the fracture surface of the composites, such as the interfacial interaction between the leather, jute fiber, and unsaturated polyester resin.

Fourier transform infrared analysis (FT-IR)

Fourier Transform Infrared Spectroscopy, also known as FT-IR Analysis or FT-IR Spectroscopy, is an analytical technique used to identify the functional group of the composite. The Fourier Transform Infrared (FTIR) spectrum of fibers was obtained using a Prestige 21 SHIMADZU spectrometer. FT-IR spectral analysis was performed within the wavenumber range of 400 to 4000 cm^{-1} with 30 scans.

Mechanical Properties of the Composites

A band saw was utilized to cut the composites materials according to the ASTM-D638-01 (dimension 60 mm ×15 mm×2 mm) method. The mechanical properties of the fabricated composites were assessed by using a universal testing machine (UTM) (Model: Testometric M-500-30 KNCT) applying a cross-head speed of 20 mm/s. All the test results were taken as the average of five values for the accuracy of the test.

Results and Discussion

Tensile properties of the composite

The mechanical properties of any material are very significant for applications for various purposes. The good mechanical properties of composites material depend on the individual constituting material. The unsaturated polyester resin was used as the matrix material having low mechanical properties. When waste jute fabrics and leather waste were added with the UPR, the mechanical properties were improved. The tensile strength of various types of composites is shown in figure 3. When the quantity of waste jute fabric was increased, the tensile property gradually decreased up to a precise limit. The height TS was for sample 4, and the value was 24 MPa. Then gradually the TS value decreases with the increment of jute percentage. A similar trend was observed for the tensile modulus value and is shown in figure 4. In this percentage, the highest value, 907 MPa was also observed for sample 4, which contained 2 % leather waste, 10% jute, and 88% UPR. Percentage elongation at break was opposite to the TS and TM values, thus gradually decreased up to sample 4. The lowest value was 0.98% and then increased.

In this research, waste jute fabrics and leather waste were strongly hydrophilic materials on the contrary UPR was hydrophobic. Thus, the bonding among the matrix and reinforcing material was not good. With the addition of jute fabrics, the hydrophobic amount was gradually decreasing. Up to a definite limit, the valance between the hydrophobic and hydrophilic was maintained say sample 4. After this valance the hydrophilic amount was increased, so the mechanical bonding was abrupt. For this reason up to sample 4, the mechanical properties were liner, and then unusual properties were witnessed. The lowest mechanical properties were measured for sample 10 containing 25% jute waste. At this percentage the composite lost its properties thus it cannot be used for any commercial purpose.

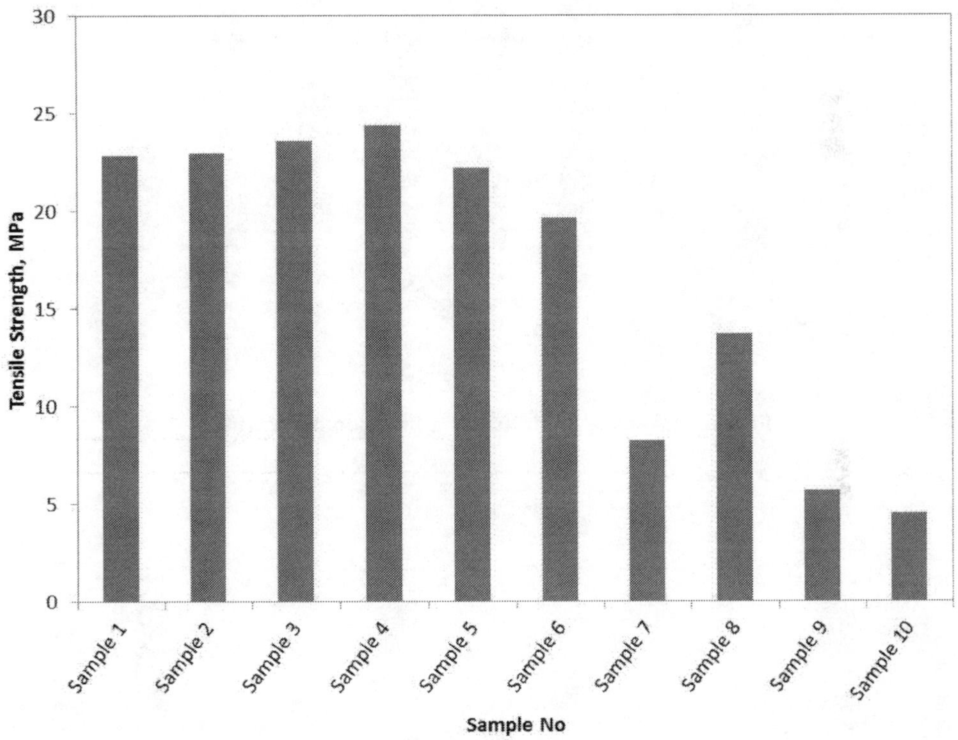

Figure 3: Tensile strength as a function of samples

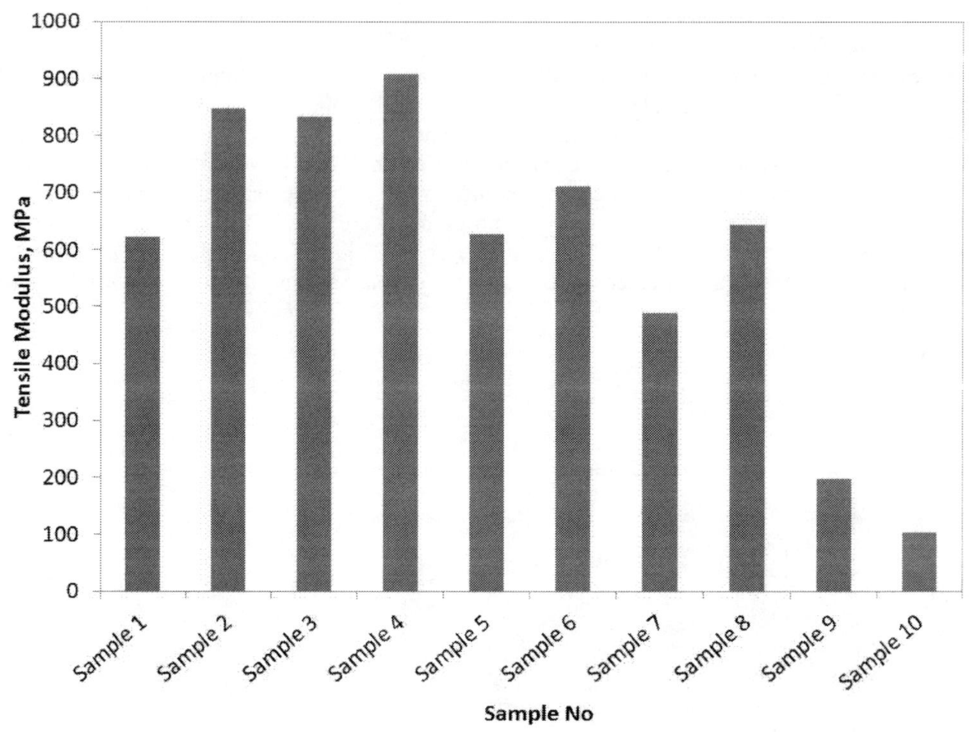

Figure 4: Tensile modulus as a function of samples

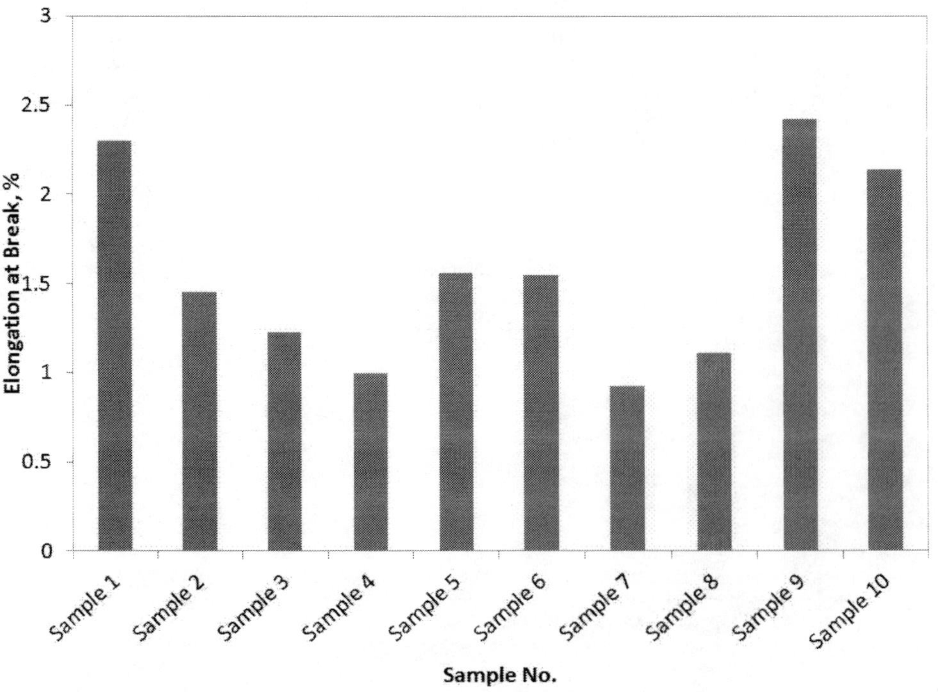

Figure 5: Percentage elongation at break as a function of samples

Scanning electron microscopic (SEM) analysis

Scanning electron microscopic (SEM) investigation of the fracture surface of the composites was performed to study interfacial properties among leather waste, fiber, and matrix. Figures 6, 7, and 8 reveal typical SEM images of natural fiber-reinforced polymer composite. In this case, fiber fracture and pull-out were noticed. In SEM images, it can be seen that the leather waste was uniformly dispersed through the matrix materials. On the contrary, the jute fabrics were not so uniform due to their long chain, rather entangled in a certain area. When the long-chain natural polymer is used as the reinforcing agent in a composites fabrication, they create void space in the composite. For this reason, it is necessary to use some additives at a certain percentage. 2% additives are best to minimize this void space [7]. In this case, waste leather of 2% carried good properties as an additive to subside the void space in composites. In SEM images no void space was noticed in the composites other than the jute fiber pull-out point. At a certain point, void space was observed and that was for the jute fiber pull-out point. When the surface of the jute filament was examined deeply, it was observed that the surface was rough. This rough surface indicated the long-time use of jute fabrics. This rough surface may be for the removal of lignin, pectin, and other binding agents as well as the cellulose breaks down from the surface due to the use of a long time. It could be observed that in all the cases the fibers were still embedded in the resin together with some cavities left by pulled-out fibers. No chemical bonding was also observed. In addition, it could be seen, there was a fiber misalignment and entanglement. Fiber alignment factors play a crucial role in the overall properties of composites. The random orientation of fibers produces lower mechanical properties compared to long unidirectional orientated fibers. This fiber entanglement can create resin-rich areas, which can contribute to the formation of voids and porosity [13]. A debonding developed between the fiber and matrix, a majority of fibers pulled out as shown in Figure 7.

Figure 6: The SEM micrograph of fractured surface of the composites

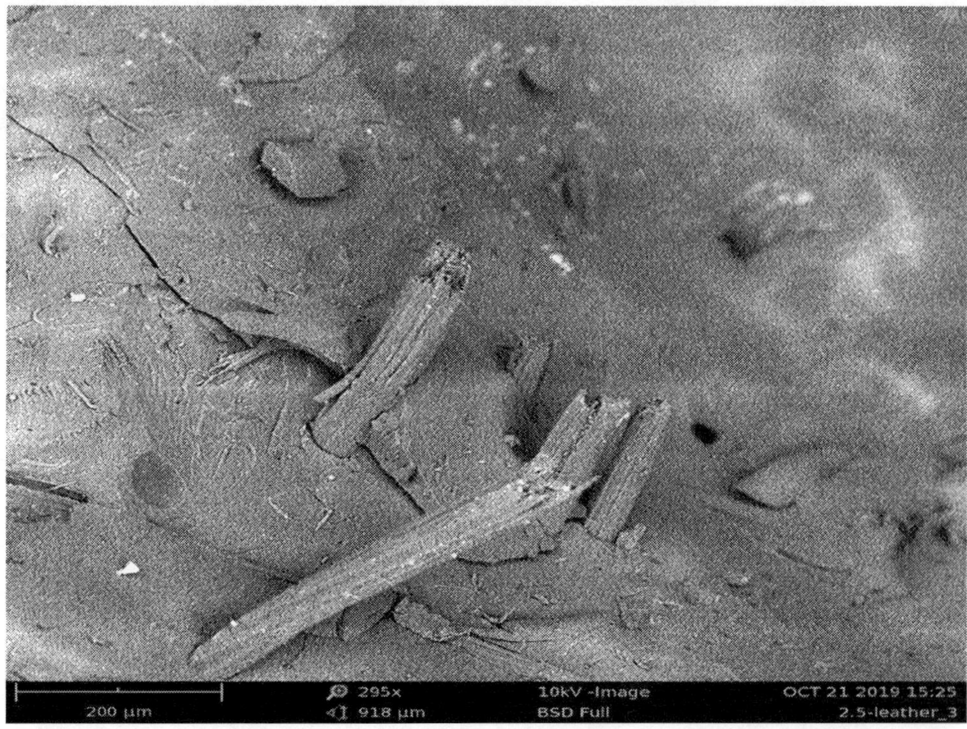

Figure 7: The SEM micrograph of fiber misalignment and entanglement

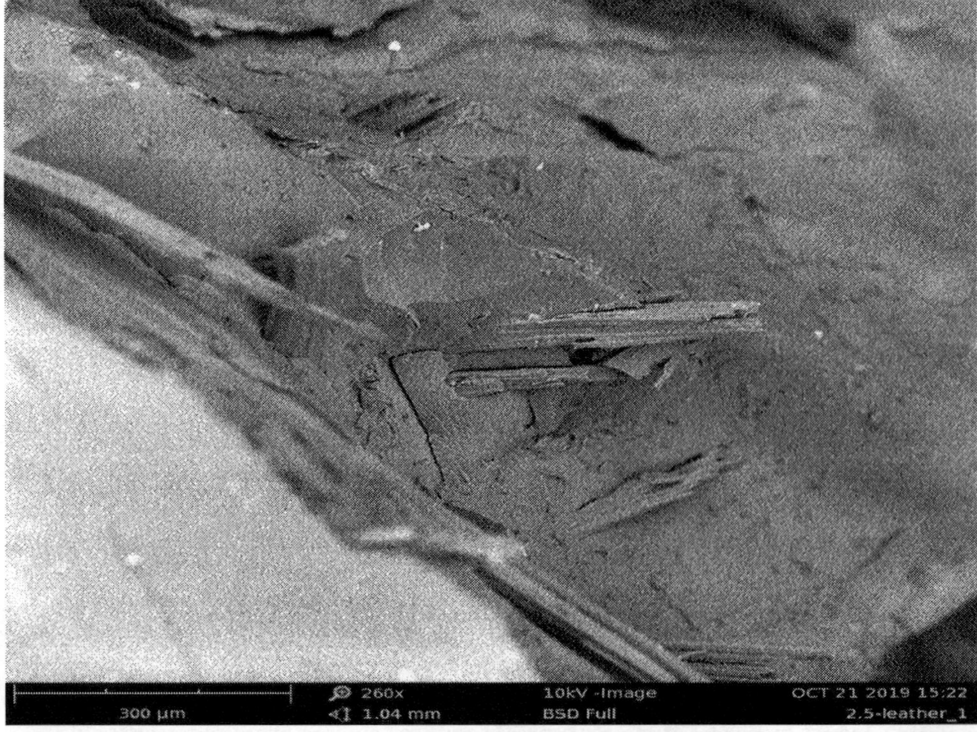

Figure 8: The SEM micrograph of good interaction of fibers and resin

Fourier transforms infrared analysis (FT-IR)

The functional groups of natural fiber reinforced polymer composites were investigated using FT-IR spectra in the range of 400 to 4000 cm^{-1}. The composite contains a constant quantity of leather besides jute and UPR. The jute fiber commonly contains cellulose, hemicelluloses, and lignin.

The principal absorption peaks in FTIR obtained for the composite are presented in figure 9. The peak around 3520 cm^{-1} was due to the stretching vibrations of hydroxyl (OH) groups. There were also different peaks observed at a wide range of wavenumber like a peak at 3061 cm^{-1} for =C-H stretching, strong peaks at 2926 and 2856 cm^{-1} for C-H stretching vibration of aliphatic methylene groups, a very intense peak at 1720 cm^{-1} due to the carbonyl (C=O) stretching from the ester linkage, medium-weak peaks at 1579, 1598 cm^{-1} due to lignin aromatic ring vibration and stretching and 1490, 1450, and 1375 cm^{-1} due to -CH, -CH$_2$ and -CH3 bending, a strong peak at 1263 cm^{-1} due to C-O-C and =C-O stretching, and lignin aromatic C=O stretching. The spectra of FTIR showed no chemical bonding among the jute fabrics, waste leather, and unsaturated polyester resin, there was only mechanical bonding.

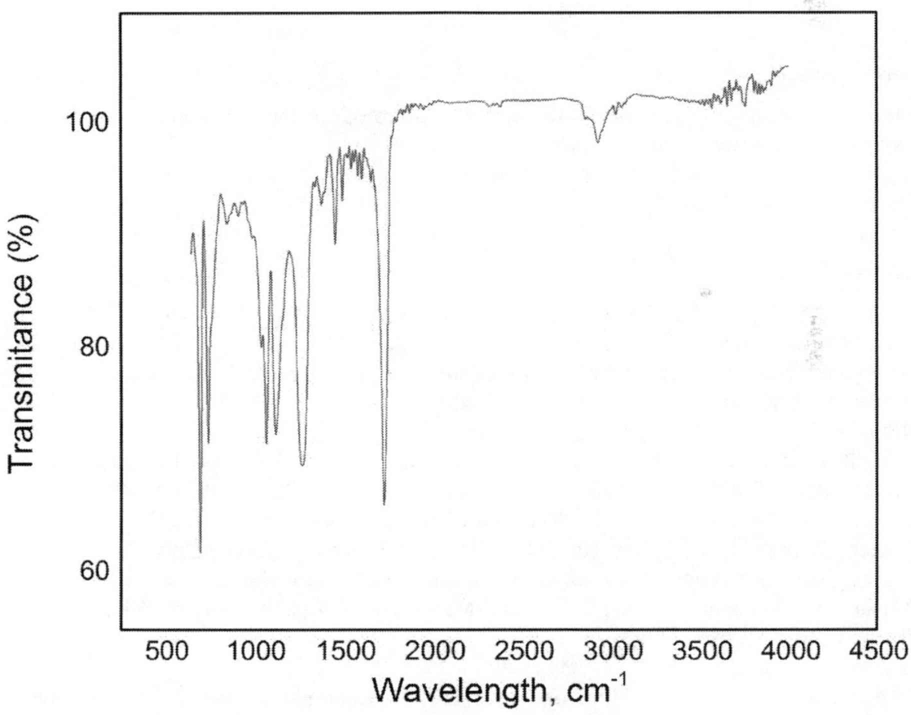

Figure 9: FTIR spectra of prepared composite

The incorporation of jute fiber or waste jute into the UPR matrix improves the mechanical properties. In addition to the jute, the precise percentage of binder also augments the mechanical properties [19, 20]. The binder not only reduces the distance between the fiber and matrix but fills the void space in the composte[21]. These are the reasons behind slightly enhanced mechanical properties. A similar pattern was noticed in our current research. No chemical bondings were detected in FTIR analysis and SEM images, so the improved mechanical properties were due to only the mechanical bondings.

Conclusion

This study focuses on the utilization of waste jute fiber and shaving wastes (cow hides) as a beneficial product in leather processing. Shaving wastes are one of the waste materials of leather industries, which contribute most to the polluting environment. Jute fibers are bio-degradable and eco-friendly natural fibers, but used jute products are creating problems. The use of jute fabrics and waste leather in composite fabrication reduces the environmental polluting agents. At the same time, the amount of synthetic polymer can be minimized which will also add benefit to the environment. For a sustainable environment, waste leather and used jute fabrics can be used to fabricate biodegradable and commercially cost-effective composite materials. So, it can be suggested from this research to use 2% leather waste, 10% waste jute fabrics in an unsaturated polyester resin matrix for the fabrication of sustainable composite materials. The composition of sample 4 will reduce environmental pollution, lower the production cost, subside the utilization of synthetic polymers, and improve the mechanical properties.

Declaration

There is no conflict of interest regarding this research.

Acknowledgement

The authors would like to give special thanks to Institute of Leather Engineering and Technology, University of Dhaka and Glass Research Division, Institute of Glass & Ceramic Research and Testing, Bangladesh Council of Scientific and Industrial Research (BCSIR) for providing required facilities for this research work.

References

[1] Sivaram, N. M., and Debabrata Barik. "Toxic waste from leather industries." *Energy from Toxic Organic Waste for Heat and Power Generation*. Woodhead Publishing, 2019. 55-67.

[2] Ravindran, B., et al. "Vermicomposting of solid waste generated from leather industries using epigeic earthworm Eisenia foetida." Applied biochemistry and biotechnology 151.2-3 (2008): 480-488.

[3] Sengül, F., and O. Gürel. "Pollution profile of leather industries; waste characterization and pretreatment of pollutants." Water Science and Technology 28.2 (1993): 87-96.

[4] Subbiah A, Chandrasekaran S, Veerasimman A. Experimental Investigations to Promote Better Adaptability of Safety Helmets Using Jute Fiber Reinforced Polyester Composites–A Waste Utilization Approach. Journal of Natural Fibers. 2021 Mar 26:1-3.

[5] Mishra, S.; Mohanty, A.K.; Drzal, L.T.; Misra, M.; Parija, S.; Nayak, S.K.; Tripathy, S.S. Compos. Sci. Technol. 2003, 63 (10), 1377–1385.

[6] Saheb, D.N.; Jog, J.P. Adv. Polym. Technol. 1999, 18 (4), 351–363.

[7]   Khan, A.R.; Haque, E.M.; Huq, T.; Khan, A.M.; Zaman, U.H.; Fatema, J.K.; Al-Mamun, M.; Khan, A.; Ahmad, A.M. J. Thermoplast. Compos. Mater. 2010, 23 (5), 665–681.

[8] Md. Sahadat Hossain, A. M. Sarwaruddin Chowdhury & Ruhul A. Khan (2017) Effect of disaccharide, gamma radiation and temperature on the physico-mechanical properties of jute fabrics reinforced unsaturated polyester resin-based composite, Radiation Effects and Defects in Solids, 172:5-6, 517-530, DOI: 10.1080/10420150.2017.1351442

[9] Shahriar Kabir, Mohammad, M. Sahadat Hossain, Monir Mia, Md Islam, Md Rahman, Mohammad Bellal Hoque, and A. M. Chowdhury. "Mechanical Properties of Gamma-Irradiated Natural Fiber Reinforced Composites." In Nano Hybrids and Composites, vol. 23, pp. 24-38. Trans Tech Publications, 2018.

[10] Rahaman, Md Niloy, Md Sahadat Hossain, Md Razzak, Muhammad B. Uddin, AM Sarwaruddin Chowdhury, and Ruhul A. Khan. "Effect of dye and temperature on the physico-mechanical properties of jute/PP and jute/LLDPE based composites." Heliyon 5, no. 6 (2019): e01753.

[11] Sahadat Hossain, M., Razzak, M., Uddin, M.B., Chowdhury, A.S. and Khan, R.A., 2019. Physico-mechanical properties of jute fiber-reinforced LDPE-based composite: effect of disaccharide (sucrose) and gamma radiation. Radiation Effects and Defects in Solids, pp.1-13

[12] Gay, D.; Hoa, S.V.; Tsai, S.W. Composite Materials: Design and Applications; CRC Press: FL, 2002; pp 203–233.

[13] Dhakal HN, Zhang ZY, Richardson MOW. Effect of water absorption on the mechanical properties of hemp fibre reinforced unsaturated polyester composites. Composites Science and Technology 2007;67:1674–83.

[14] Ahmadi R, Alivand MS, Tehrani NH, Ardjmand M, Rashidi A, Rafizadeh M, Seif A, Mollakazemi F, Noorpoor Z, Rudd J. Preparation of fiber-like nanoporous carbon from jute thread waste for superior CO_2 and H_2S removal from natural gas: Experimental and DFT study. Chemical Engineering Journal. 2021 Jul 1;415:129076.

[15] Ahmadi R, Ardjmand M, Rashidi A, Rafizadeh M. Enhanced gas adsorption using an effective nanoadsorbent with high surface area based on waste jute as cellulose fiber. Biomass Conversion and Biorefinery. 2021 Apr 29:1-6.

[16] Kamble Z, Behera BK. Sustainable hybrid composites reinforced with textile waste for construction and building applications. Construction and Building Materials. 2021 May 17;284:122800.

[17] Singh CP, Patel RV, Hasan MF, Yadav A, Kumar V, Kumar A. Fabrication and evaluation of physical and mechanical properties of jute and coconut coir reinforced polymer matrix composite. Materials Today: Proceedings. 2021 Jan 1;38:2572-7.

[18] Patti A, Cicala G, Acierno D. Eco-Sustainability of the Textile Production: Waste Recovery and Current Recycling in the Composites World. Polymers. 2021 Jan;13(1):134.

[19] Hossain MS. Studies on the degradation of jute blended cotton fabric in soil, water and ambient atmosphere. International Journal of Environmental Chemistry. 2018 Nov 27; 4(2): 1-1.

[20] Ali MF, Hossain MS, Ahmed S, Chowdhury AS. Fabrication and characterization of eco-friendly composite materials from natural animal fibers. Heliyon. 2021 May 1;7(5):e06954.

[21] Hossain MT, Hossain MS, Uddin MB, Khan RA, Chowdhury AS. Preparation and characterization of sodium silicate–treated jute-cotton blended polymer–reinforced UPR-based composite: effect of γ-radiation. Advanced Composites and Hybrid Materials. 2020 Jul 14:1-8.

Nano Hybrids and Composites
ISSN: 2297-3370, Vol. 33, pp 13-34
© 2021 Trans Tech Publications Ltd, Switzerland

Submitted: 2021-06-06
Revised: 2021-08-27
Accepted: 2021-08-31
Online: 2021-10-11

The effect of Nanoparticles (n-HAp, n-TiO$_2$) on the Thermal Properties and Biomechanical Analysis of Polymeric Composite Materials for Dental Applications

Alaa A. Mohammed[1,a*], Jawad K. Oleiwi[2,b] and Emad S. Al-Hassani[3,c]

[1,2,3]Department of Materials Engineering, University of Technology –Iraq

[a]alaa.a.mohammed@uotechnology.edu.iq, [b]130041@uotechnology.edu.iq, [c]130042@uotechnology.edu.iq

Keywords: PEEK, n-HAp, n-TiO$_2$, Thermal conductivity, DSC, Finite element method.

Abstract. Polyetheretherketone (PEEK), as implants is broadly employed in orthopedic and dental uses owing to the brilliant chemical stability, biocompatibility and mechanical strength in addition to the modulus of elasticity alike the human bone. In the present work, the composite materials with PEEK as matrix and nanohydroxyaptite (n-HAp), nanotitaniumdioxide (n-TiO$_2$) as the reinforced fillers added with weight fractions (0, 0.5, 1 and 1.5 wt%) were prepared by internal mixer and hot press. Following analysis by physical properties includes the thermal conductivity and the differential scanning calorimetry. Finite element analysis (FEA) was used to find the total deformation, Max. Von mises stress, elastic strain and safety factor. The results manifested that the thermal properties, total deformation and strain decreased with the increase of the reinforcement weight fraction, while, the stress and safety factor increased with the increased reinforcement weight fraction.

1. Introduction

Metals and alloys of medical grades were presented into biomedical area from the too beginning owing to their high performance and are broadly utilized for clinical implants currently, for instant, titanium (Ti) and its alloys are often selected as the materials for orthopedic and dental implants [1, 2]. Despite the preferable performance of wear, low toxicity, good formability, and the high strength are the whole favorable characteristics that metals can offer [1], remarkable drawbacks still present, for example, the artifacts in the X-ray, the high modulus of elasticity, the restriction in the imaging of the magnetic resonance, the release of the prospective metal ion, and the consequent allergenicity or osteolysis, which cannot be ignored in the clinical events [3, 4]. Most significantly, the influence of the stress shielding resulted via the high modulus of elasticity is usually a prospective danger for the metal implants [5, 6].

Presently, thermoplastics are encouraging replacements for being utilized as structural materials in the biomedical uses for replacing the Ti-based and ceramic materials that have been utilized for some decades [7]. Poly(oxy-1,4-phenyleneoxy-1,4-phenylenecarbonyl-1,4-phenylene), highly recognized as a polyether-ether-ketene (PEEK) is a totally aromatic semi-crystalline thermoplastic engineering plastic. The macro-molecular chain includes a rigid benzene ring, stretchy ether bonds and carbonyl sets that enhance the inter-molecular interactions, providing such high performance polymer with high modulus, high fracture toughness, high strength, and other brilliant complete performance [8]. Distant from that, it also could maintain its mechanical structure in the harsh surroundings, because PEEK possesses a high thermal stability and a high chemical resistance [9].

Hydroxyapatite (HA), [Ca$_{10}$(PO$_4$)$_6$(OH)$_2$], is a calcium phosphate mainly known for its applications in bone replacement [10]. In recent years, the hydroxyapatite reinforced polymer composites have taken much interest, particularly in the biomedical materials field. Dissimilar to the traditional implant materials, like Co-Cr alloys and stainless steel alloys, the following biomaterials generation (for example HA reinforced polymer composites), with the bioactive HA inclusion not only corresponds the mechanical properties (i.e. strength and stiffness) of bon, but also mimics the bony tissue via making a biological reaction that encourages the growth of bone upon the implant [11].

Titanium dioxide TiO_2 is well known as biologically and chemically inert, mechanically robust, nontoxic, cheap, biocompatible, antifogging and super-hydrophobic [12]. TiO_2 is broadly utilized in the ecological uses, paper, paints, foods, cosmetics, toothpastes, and coatings, due to its green, unpolluted, less cost and maintainable innovation also frequently utilized as a polymer matrix reinforcement phase to enhance its mechanical properties [13].

Finite element analysis is a powerful tool for designing has been effectively utilized for assessing the design property of the mostly intricate mechanical assembly geometries. It's a technique to break down an intricate geometry into handy "elements" or simpler geometries that their boundary circumstances are recognized. The solid models being utilized as a foundation for the finite element analysis. Constructing the solid models is the initial stage, where they are able to more transfer straight into the FA software to execute the finite element analysis. Owing to the truth that the method of the finite element analysis possesses the capability to treat with the intricate geometry, the intricate behavior of material, and the intricate loading circumstances that commonly met in the uses of the dental implants, a bigger number of the FEA investigations were performed to study the deigns of the dental implant [14].

Schwitalla et al. illustrated the FEA for pointing out the discrepancies in the biomechanical trend of a dental implant of Endoligns and a commercial powder-filled polyetheretherketone. Ti acted as a control. The whole kinds revealed a minimum safety factor concerning the cortical bone yield strength. Nevertheless, within this investigation limits, the Type 2 implant depicted higher stresses within the near cortical bone than Type 1 and Type 3. Such implant assemblies exhibited the same distributions of stress [15].

Tang et al. investigated the cyclic load upon the PEEK/HA composite with various HA contents and manifested that the HA/PEEK composite is an encouraging fatigue-resistant material for the biomedical uses [16].

Wong et al. illustrated that the bending modulus of the strontium-containing hydroxyapatite/polyether ketone Sr-HA/PEEK rose with the increasing of the Sr-HA volume fraction. The elastic modulus of (25 vol%) and (30 vol%) Sr-HA reinforcement revealed (113%) and (136%) increment, correspondingly compared with the neat (PEEK). The bending strengths of (25 vol%) and (30 vol%) Sr-HA reinforcement demonstrated (25%) and (29%) reduction, correspondingly compared with the neat (PEEK) [17].

Kim et al. integrated the $30CaO \cdot 70SiO_2$ (CS) microspheres with the PEEK for obtaining a bioactive bone-repairing material with mechanical characteristics same as those of the natural bone. It was noticed that the hydroxyapatite development upon the whole CS-PEEK composites surfaces beyond the immersion into the simulated body fluid SBF, demonstrating a potential bond to the natural bone. In addition, the bending strength and stiffness rose in 6% and in 38%, correspondingly with the (20 vol% CS) particles addition. An important reduction of the composites mechanical properties beyond the immersion in SBF was not noted. [18].

Wang et al. combined the nano FHA with the PEEK to improve its antibacterial property and the implant materials osseointegration. The mechanical tests elucidated that the nano FHA works as a reinforcement filler in the PEEK regarding that the modulus of elasticity and the tensile strength raised in (210%) and (67%), correspondingly, with the (40 wt%) nano FHA incorporation into the (PEEK). Moreover, the microbiological assays indicated that the composite possesses an antibacterial activity. The in-vivo assays evinced the bone-implant contact, the high bioactivity, and osseointegration [19].

Oliveira et al. enhanced the osseiontegration of the injection molded PEEK, plasma sprayed hydroxyapatite (HA) coatings are used using three distinct thermal conditions for PEEK: (i) as molded, (ii) plasma-exposed only, and (iii) HA-coated [20].

Masamoto et al. evolved a surface processing called PrA (Precursor of Apatit) for the polyetheretherketone (PrA-PEEK) by a simple, low-temperature procedure that aims to perform a vigorous and quicker adhesion to the bone. This processing includes (3) stages: The immersion of the sulfuric acid (H_2SO_4), the (O_2) plasma discharge exposure, and alkaline simulated body fluid (alkaline SBF) processing. The in-vitro investigation utilizing the (MC3T3-E1) cells depicted by the

XTT assay that the (PrA) upon the substrate of the PEEK caused no cytotoxicity, throughout the PrA processing appeared to suppress the integrin gene expression beyond the (7) days of incubation [21].

Bataineh et al. investigated the effect of using carbon reinforced PEEK composite material for fixture/abutment on the stress distribution in periimplant bone by using 3D FEA. The results showed that there is no significant difference in the distribution pattern of stress at the implant-bone interface among the different material models studied. The highest maximum and the lowest minimum principal stresses were always located in the cortical bone and never in the cancellous bone which is consistent with the existing literature [14].

Lv et al. improved the mechanical, tribological and biological performances of the polyetheretherketone (PEEK) by incorporating molybdenum disulfide (MoS_2, MS) nano sheets into the MS content of 4 wt% and 8 wt% [22].

The main purpose of this study is to determine the thermal properties of a novel composite materials used as dental implants made by PEEK reinforced with n-HAp and n-TiO_2 separately at different weight fractions (0, 0.5, 1, and 1.5%) for two types of reinforcement. A thermal test was conducted in order to evaluate the thermal conductivity for the prepared composites. Also, a differential scanning calorimetry test was performed to illustrate the glass transition temperature, melting temperature, crystalline temperature and the degree of crystallinity value for the resultant composites. Moreover, Finite Element Analysis was done to investigate the maximum Von Mises stress, total deformation, elastic strain and the safety factor of the polymer composites. The novelty of the present study is manufacturing of composite materials by compounding and hot press process. In addition to use ANSYS program to predict the behavior of prepared composites under certain conditions.

2.   Materials and Methods

2.1 Raw Materials

Polyetheretherketone (551G) granule, nano hydroxyapatite and nano titanium dioxide were purchased from the (Jilin Joinature Polymer Co., Ltd, China), (N and R Industries, Inc., China) and (Hangzhou Union Biotechnology Co., Ltd, China), correspondingly. The (TiO_2) and (HAp) nano particles size was 20 nm.

2.2 Preparation of Samples

The biopolymer composites containing (0, 0.5, 1, and 1.5) wt% for the two types of nano powder were fabricated via a series of processes compounding and hot press process, as method in reference [23, 24]. PEEK granule was desiccated in an oven at 80°C for one hour prior to the process of compounding and hot press. Briefly, the nano powders were distributed in the ethanol alcohol utilizing an ultrasonic blender with variables (90 W power, 0.5 pulse and (5-10) min) to obtain a homogeneous mixture and then mixing with PEEK granule. After well dispersed, the mixture was dried in an oven at 90°C for 24 hour to remove the excess ethanol alcohol.

In the compounding process, the compounding was carried out in the an internal blender (HAAKE) type (Hbisystem 90, Ahaake Buchler Product, USA) at 360°C and a blending speed of (90) rpm. The time needed for compounding was adjusted accordingly to 20 min. Then, the hot press by using mold with dimension (17*17*0.4) cm^3 was preheated at temperature 200°C and covered from the top and bottom by a cover to withstand a high temperature in order to prevent the adhesion of the polymeric composite to mold and put in hot press type (Toyoseiki, Japan). The hot press temperature was adapted at (360°C) and when reaching this temperature, the pressure was exerted for 15 minutes and its value is equal to (15 MPa). Then, the mold was taken out from the hydraulic press and put in a system of cooling by water jet (5 L/min) at the room temperature. The mold was opened, then the composite sheet was removed, as shown in Fig.1, and the specimens were cut by CNC machine for each test.

Fig.1 Some of prepared samples

2.3 Characterization

The thermal conductivity of a material is a measure of its ability to conduct heat. It is commonly denoted by Lees disk commonly used to determine. This test was performed on (FARNELL INSTRUMENTS LTD., Britain). Differential scanning calorimetry was used in order to study the thermal behavior of the prepared polymer composite samples according to ASTM D3418-03 [25]. The measurements of the differential scanning calorimetry (DSC) were performed using a DSC (METTLER TOLEDO (Switzerland)). An amount of material of almost (10–15) mg was airtight in a standard aluminum crucible in nitrogen atmosphere. Samples were massed and heated from 25°C to 400°C at a rate 10°C/min. The mechanical tests involve tensile test was performed on the tensile test machine (SANTAM, type STM-50, Iran). Also nanoindentaion test was performed on (Triboscope, nanomechanical test instrument , HYSITRON, Inc, USA).

3. Steps of Finite Element Method

The single steps involved in the analysis will be discussed sequentially; these are preprocessing (i.e., geometrical modeling and meshing), solution of the discretized equations, and post processing. For illustration reasons, ANSYS was used, which represents one of the first commercial FEM codes available [26].

3.1 Geometric Modeling

The first step in FEM is a geometric modeling. In this step, by using the ANSYS program, a model of axisymmetric finite element of screw dental implant is built according to the dimensions of it, as shown in Fig. 2.

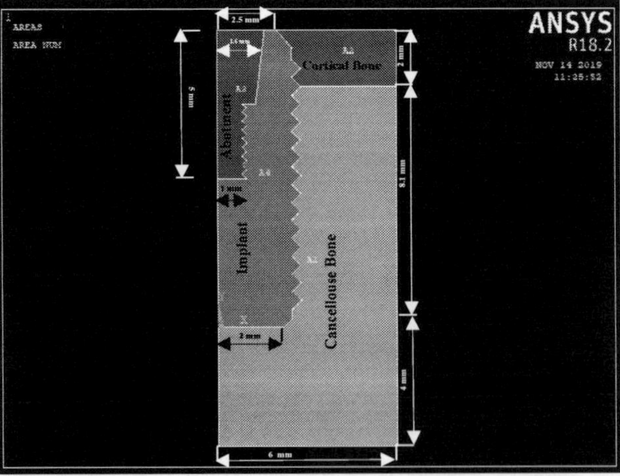

Fig.2 Fixture of prepared implants in the bone

3.2 Element Type

Two dimensional solid structures (PLANE182) by (ANSYS-18.2) is the element type chosen for this work. This element is utilized as either an axisymmetric element or a plane element (plane strain, generalized plane strain or plane stress). It's characterized via (4) nodes with (2) degrees of freedom at every node (translations in the nodal x and y directions). It possesses stress stiffening, plasticity, hyper-elasticity, and big strain abilities, and big deflection [27].

3.3 Material Properties with Mesh

The majority of the element kinds need material properties relying upon the use of these materials. In this study, the assumption was made that the materials were homogenous, axisymmetric, linear and had elastic behavior characterized by Young's modulus and Poisson's ratio. Young's modulus was determined experimentally from the nano indentation, which was illustrated in a previous work [28], while the samples' Poisson's ratio was theoretically computed from the mixture rule. On the other hand, the Young's modulus and Poisson's ratio for bone are illustrated in Table 1 [29].

Table 1. Properties of bone

Type	Young's modulus(GPa)	Poisson's ratio
Cortical bone	14	0.30
Cancellous bone	1.37	0.31

Meshing involves discretization of geometric model in to finite discrete elements. Meshing the screw implant is automatically generated by using meshing tool in (ANSYS-18.2), as shown in Fig. 3.

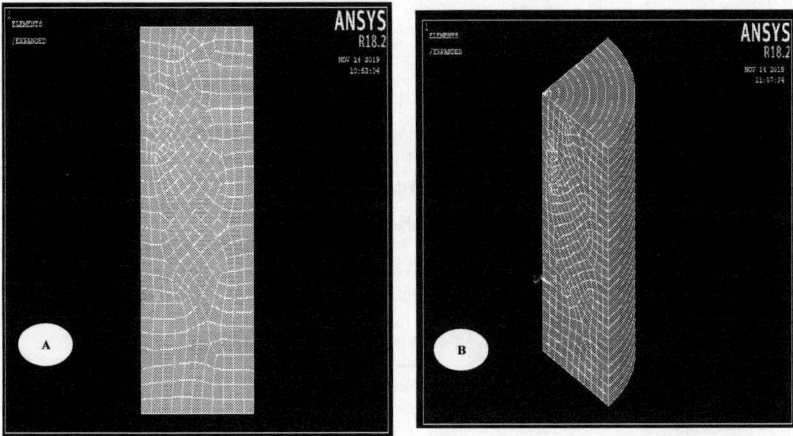

Fig.3 The mesh of samples implant A) front view b) 3D axisymmetric

3.4 Loads and Boundary Conditions

Loads in the ANSYS terminology comprises the boundary condition and the functions of the internally or externally exerted force, like the displacements, forces, pressure, gravity in structural disciplines, as shown in Fig.4. In this work, load was applied as pressure on the top of Abutment of screw with (12.434 MPa) = (100N) [30].

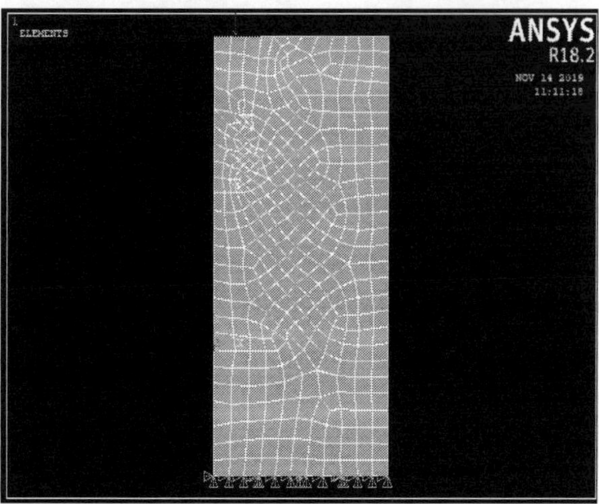

Fig.4 Loads and boundary conditions applied

3.5 Analysis and Obtaining the Results

The solution of equations is a standard procedure in matrix algebra, usually based on the Gaussian–elimination process. The steps used to solve of the Finite element analysis of the dental implant in bone. Several ways are available to obtain the results from the (ANSYS-18.2) package program. The Von Mises stress is a geometrical amalgamation of the whole stresses (the normal stress in three directions and three shear stresses) performing at a specific position and permitting the most complicated stress situation to be represented by a single quantity. Von Mises stresses can be calculated by the following equation (1), and the safety factor equation can be written in accordance with Von Mises (Distortion Energy) failure theory and calculated from equation (2).

In addition, the Von Mises failure theory was used to determine the equivalent elastic strain and the total deformation distribution values [31, 32].

$$\sigma_{vm} = \sqrt{\frac{(\sigma_x-\sigma_y)^2 + (\sigma_y-\sigma_z)^2 + (\sigma_z-\sigma_x)^2 + 6(\tau_{xy}^2+\tau_{yz}^2+\tau_{zx}^2)}{2}} \qquad (1)$$

Where:

σ_x = The principal normal stress in x-dircetion (MPa).
σ_y = The principal normal stress in y-direction (MPa).
σ_z = The principal normal stress in z-direction (MPa).
τ_{xy} = The shear stress in xy-plane (MPa).
τ_{yz} = The shear stress in yz-plane (MPa).
τ_{zx} = The shear stress in zx-plane (MPa).

$$n = \frac{\sigma_y\,(\sigma_u)}{\sigma_{vm}} \qquad (2)$$

It must be $\qquad\qquad \sigma_{vm} \leq \sigma_y$ (safe).

Where:

n = Safety factor.
σ_y = Yield stress.
σ_u = Ultimate tensile stress.
σ_{eq} = Equivalent stress.

3.6 Statistical Analysis

Thermal conductivity results were performed at descriptive and one-way analysis of variance (ANOVA) using SPSS 22.0 statistical software that was used to determine the statistical significance of the differences. The (P) value (< 0.05) was regarded to be statistically significant.

4. Results and Discussion

The average particle diameter value for n-HAp and n-TiO$_2$ was (81.87 and 95.77) nm, roughness average was (19.2 and 8.5) nm, particle size distribution and surface topography for was determined by Atomic Force Microscope (AFM) device, as show in Fig. 5 and Fig. 6.

a)

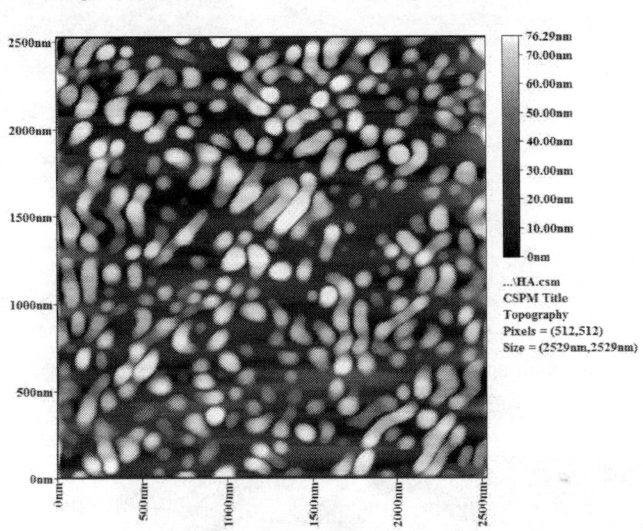

b)

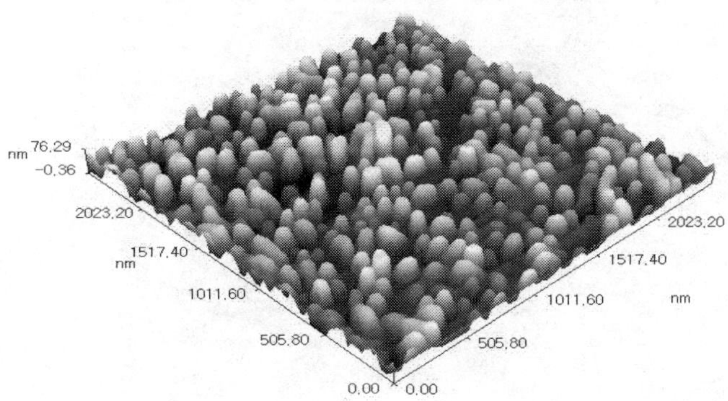

c)

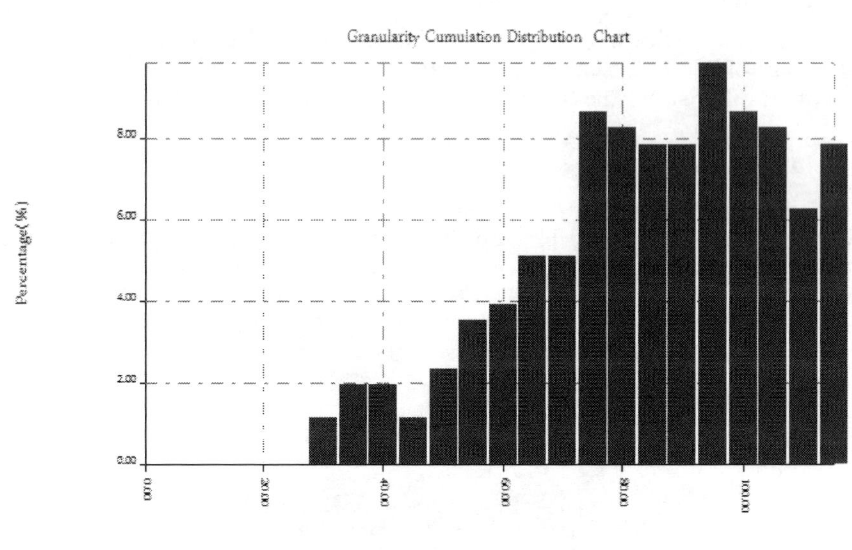

Fig. 5 AFM results for Nano hydroxyapatite: a) 3D image b) 2D image c) grain distribution

a)

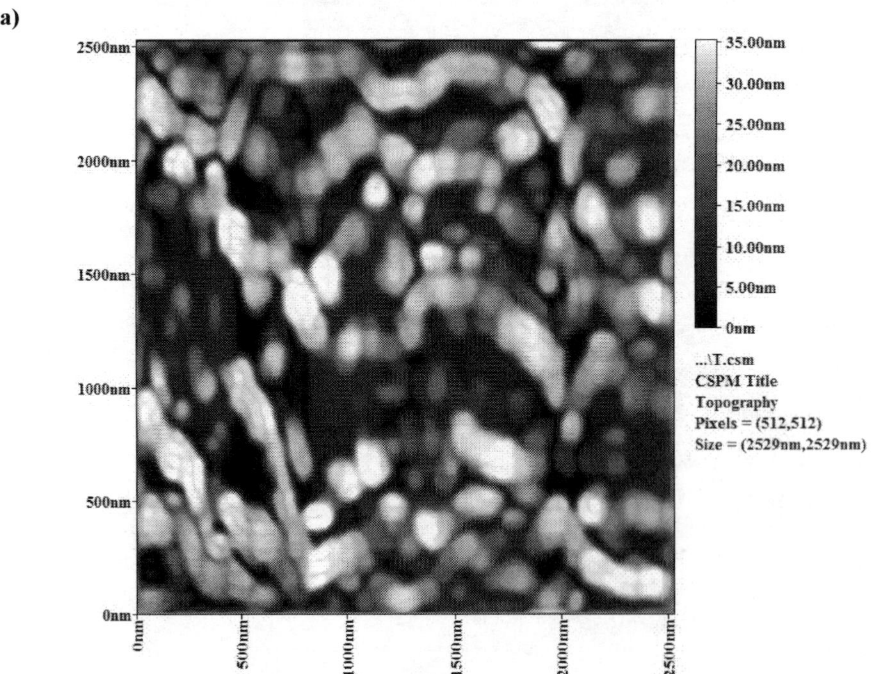

b)

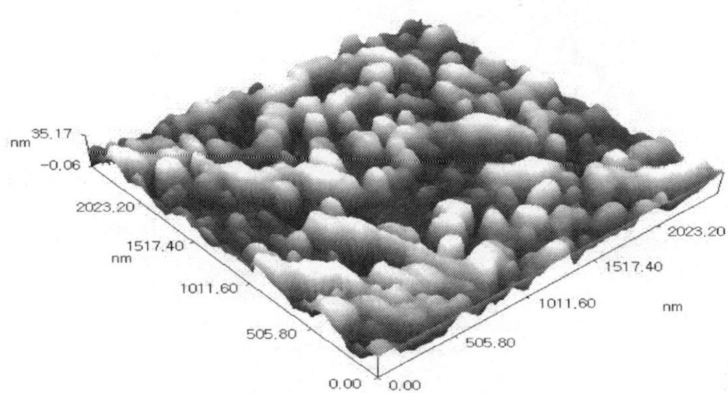

c)

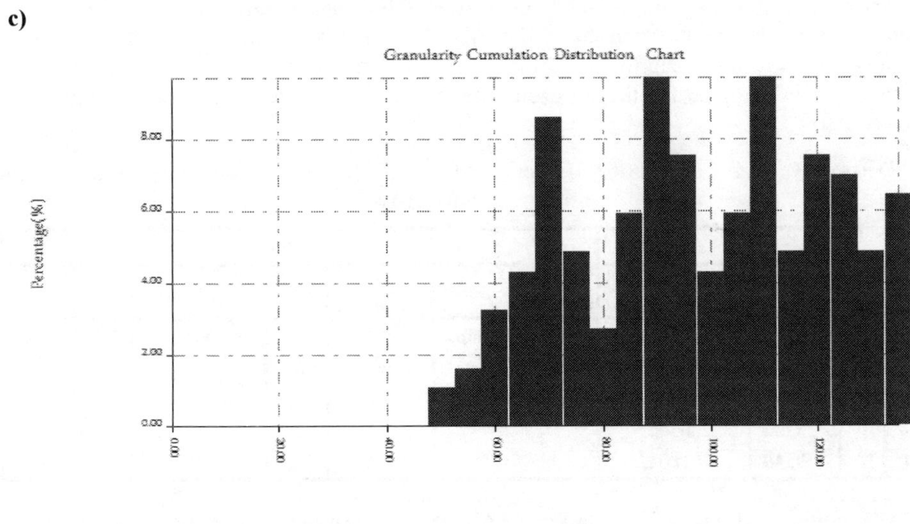

Fig. 6 AFM results for Nano titanium dioxide: a) 3D image b) 2D image c) grain distribution

4.1 Thermal Conductivity Results

The prepared composites with varying types and weight fractions of nano reinforcement were tested to determine their thermal conductivity. The results observed in Fig.7 revealed that the thermal conductivity decreases with increasing the weight fraction of nano particles in composite materials compared to the thermal conductivity of the PEEK matrix. The reason behind such behavior is the increased thermal conduction resistance with the increased weight fraction of nano particles, and this resistance increased by using nano particles sizes [33].

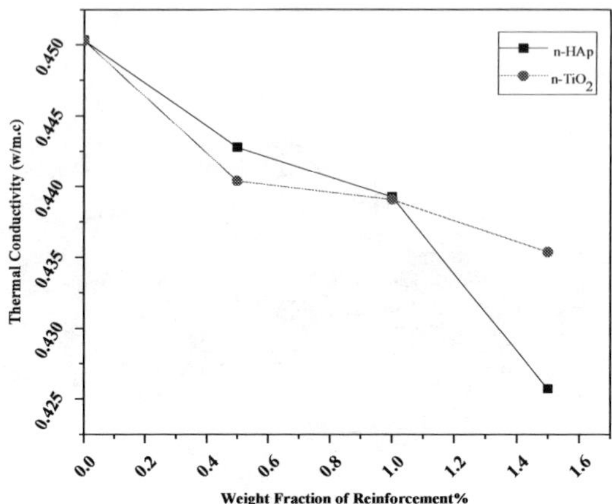

Fig.7 Thermal conductivity results for polymer composite

Tables 2 and 3 explain the statistical analysis results, for the descriptive analysis the highest mean values were obtained at a neat sample, while the lowest value was obtained at 1.5% for two types of reinforcement. Also, it can be seen from the ANOVA analysis that the significant difference between the groups was a non-significant difference for the all samples ($P \geq 0.05$). This means that the reinforcement in samples for the all cases has a negative effect on the thermal conductivity property.

Table 2 Descriptive and Analysis of Variance (ANOVA) of thermal conductivity for the samples reinforced with HAp

Descriptive				ANOVA					
Groups	N	Mean	Std. Deviation		Sum of Squares	df	Mean Square	F	Sig.
0 %	3	.450330	.2108331	Between Groups	.001	3	.00	.003	1.000
0.5 %	3	.442800	.4694249	Within Groups	1.001	8	.125		
1 %	3	.439333	.4121290	Total	1.002	11			
1.5 %	3	.425720	.2565627						
Total	12	.439546	.3018013						

Table 3 Descriptive and Analysis of Variance (ANOVA) of thermal conductivity for the samples reinforced with TiO$_2$

Descriptive				ANOVA					
Groups	N	Mean	Std. Deviation		Sum of Squares	df	Mean Square	F	Sig.
0%	3	.450330	.2108331	Between Groups	.000	3	.000	.001	1.000
0.5%	3	.440400	.5625495	Within Groups	1.163	8	.145		(NS)
1%	3	.439100	.3120351	Total	1.163	11			
1.5%	3	.435410	.3510398						
Total	12	.441310	.3252105						

4.2 Differential Scanning Calorimetry (DSC) Results

The thermal transition of the polymer samples was determined by (DSC). The important thermal transitions include the glass transition temperature (T_g), the crystallization temperature (T_c), and the melting temperature (T_m), as depicted in the Fig. 8.

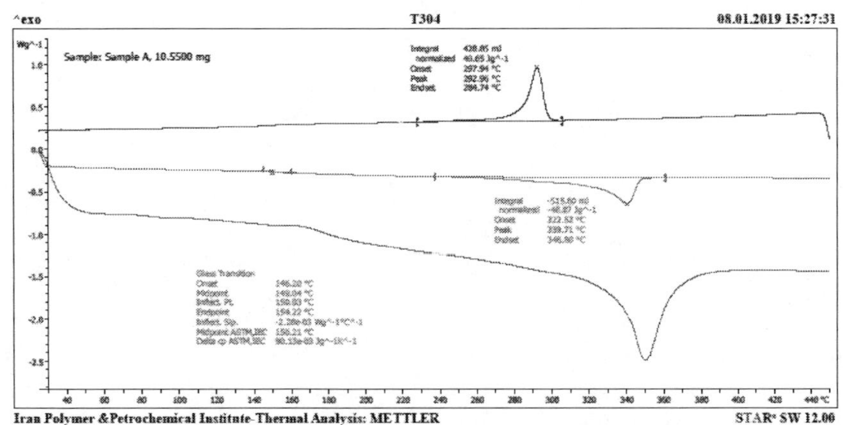

(a)

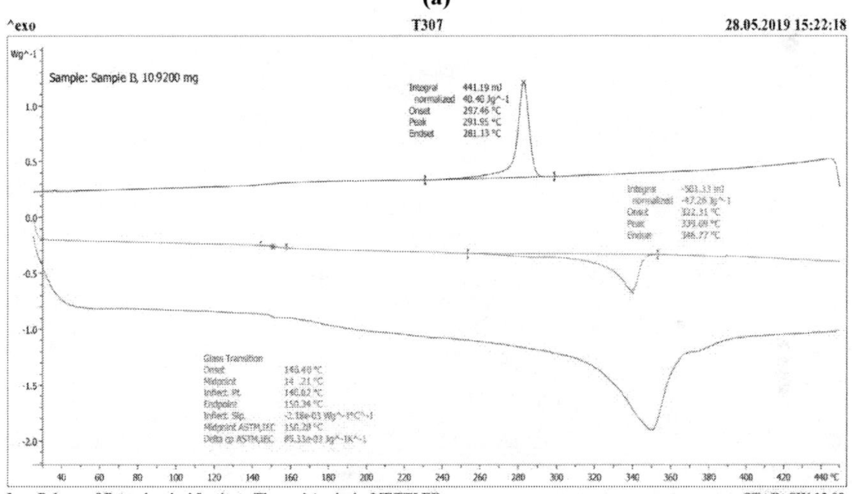

(b)

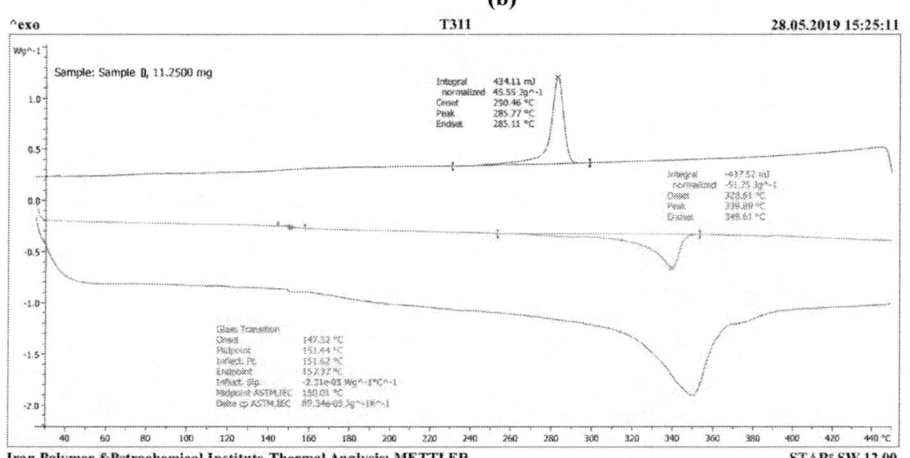

(c)

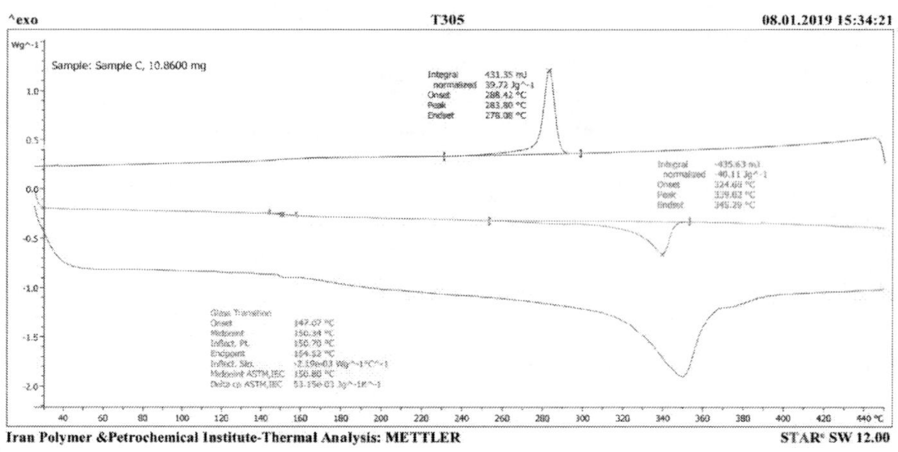

(d)

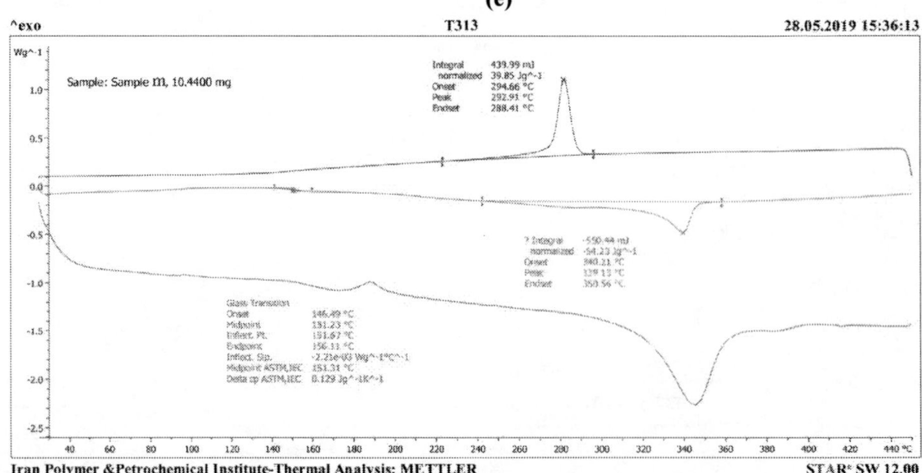

(e)

(f)

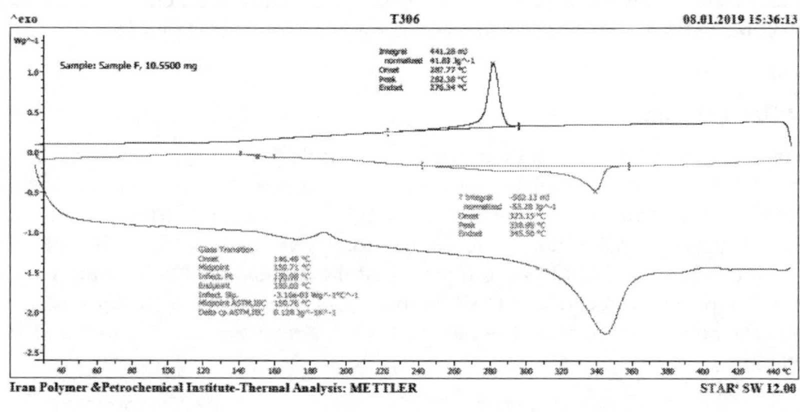

(g)

Fig.8 Thermal analysis results for a) neat PEEK b)PEEK/0.5%HAp c) PEEK/1%HAp d) PEEK/1.5%HAp e)PEEK/0.5%TiO$_2$ f) PEEK/1% TiO$_2$ g) PEEK/1.5% TiO$_2$

The temperature of melting (T_m), the temperature of crystallization (T_c), the temperature of the glass transition (T_g) and the crystallization percentage ($Xc\%$) are given in the Table (4). The crystallization percentage ($Xc\%$) was computed by this formula (3) [34]:

$$X_C\% = \frac{\Delta H}{\Delta H^{\circ}_m * W_{PEEK}} * 100\% \tag{3}$$

Where,
ΔH and ΔH°_m: Values of the fusion enthalpy of the (PEEK) based composites and the completely crystalline (PEEK), i.e., (130 J/g), respectively
W_{PEEK}: The PEEK weight fraction in the composites.

Table 4 The temperature of melting (T_m), the temperature of the glass transition (T_g), the temperature of crystallization (T_c), and the crystallization percentage ($X_c\%$) for the (PEEK) polymer composites

Sample	T_g	T_c	T_m	$X_c\%$
PEEK	146.20	292.96	339.71	37.59
0.5% HAp	146.40	291.95	339.69	39.37
1% HAp	147.32	285.77	339.89	40.21
1.5% HAp	147.07	283.80	339.02	31.32
0.5% TiO$_2$	146.25	293.21	339.06	41.35
1% TiO$_2$	148.80	283.54	339.99	42.35
1.5% TiO$_2$	146.49	282.38	338.99	41.61

The melting temperature (T_m) and the glass transition temperature (T_g) values for all samples didn't change basically and were relatively constant (~339°C and ~146°C, correspondingly), thus showing that the nano particles addition possesses no different influence upon the polymer T_m and T_g. Therefore, the PEEK crystal size doesn't change after the addition of nanoparticles. The T_c of the composites showed a decreasing trend due to the inhibitory effect of the reinforcement on the PEEK mobility [35, 36]. On the other hand, as n-TiO$_2$% and n-HAp% increase from 0.5% to 1.5%, Xc% peaked exhibits somewhat increases or decreases, i.e. a similar level to that of pristine PEEK (37.59%). This indicates that 1% nano particles has the strongest nucleation effect that promotes the formation of the microcrystalline zones within the composite [37], whereas further increasing nano particles would lead to the formation of more imperfect crystals [38]. It should be noted that the PEEK, as a semi-crystalline polymer, consists of both crystalline and amorphous regions. A lower

Xc% can be associated with the greater mobility of the polymer segments within the amorphous region, which contributes to the enhancement of the polymer toughness [39-41].

4.3 Numerical Results

4.3.1   Total Deformation Results

Fig. 9 views the contours plots of total deformation of samples for all specimens that are prepared in this study. Fig. 10 clarifies the relationship between the results of the total deformation and the type of samples. One can see from this figure that the total deformation distribution that occurs in the types of samples which explain the value of the total deformation depends upon the type of material. Also, it can be noticed from such figures that the highest total deformation value occurred in the pure PEEK specimen is equal to (13.408 mm). But, the lowest total deformation value that can be noted in the polymer specimens containing 1.5% HAp particles added to the PEEK matrix is equal to (9.263 mm).

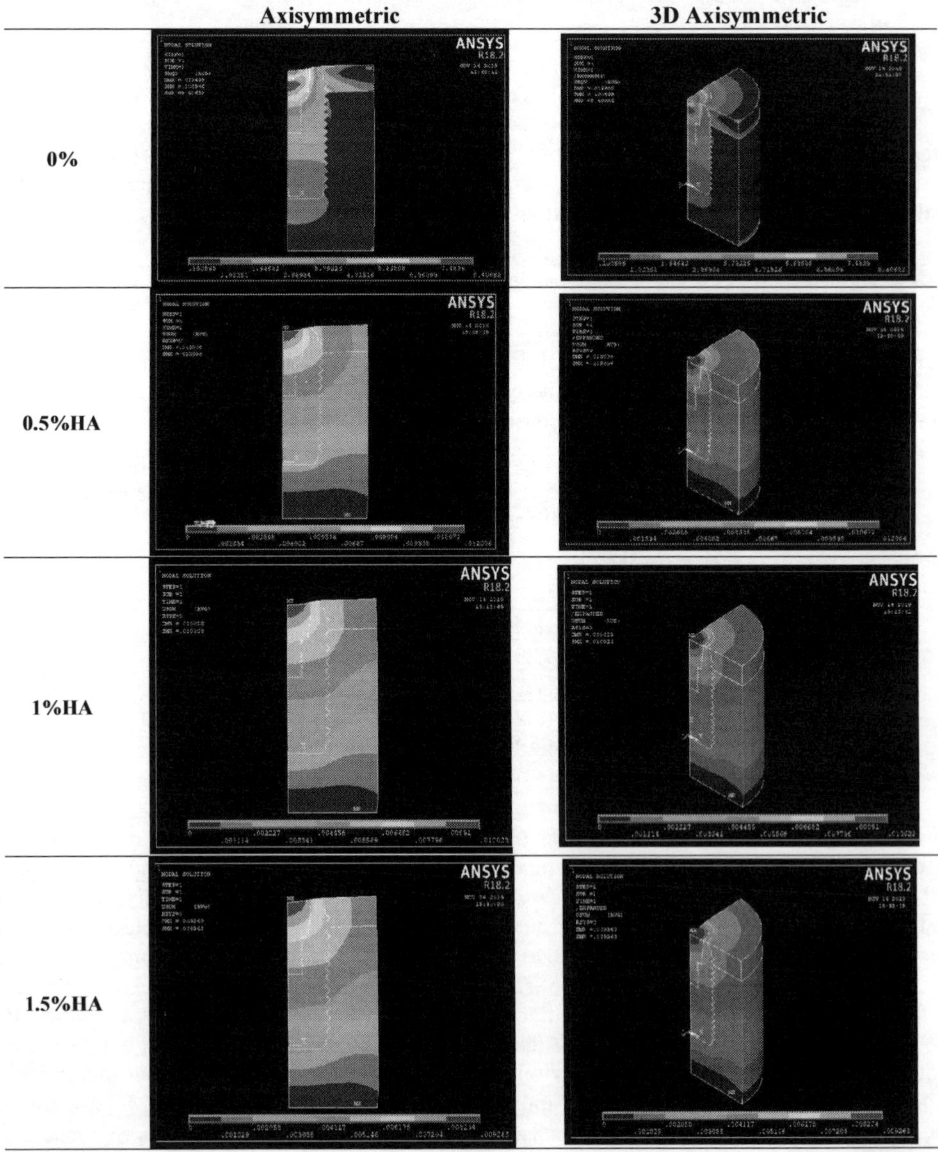

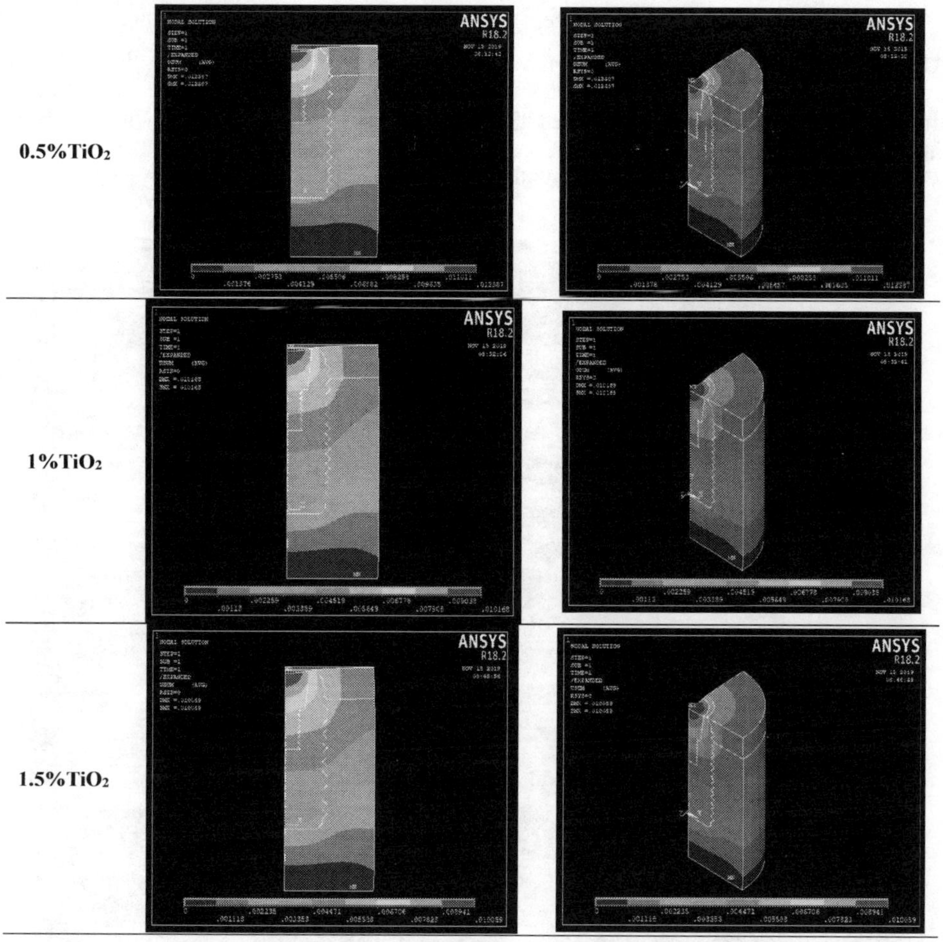

0.5%TiO₂

1%TiO₂

1.5%TiO₂

Fig. 9 Deformation result of the finite element analysis for the polymer composites

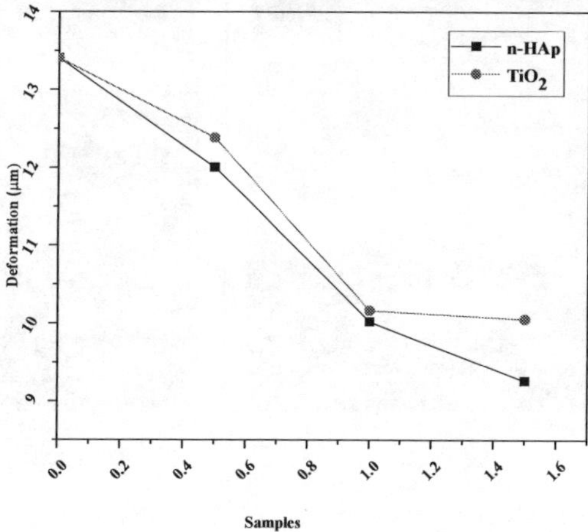

Fig.10 Total deformation results for the polymer composites

4.3.2   Equivalent Stress (Von Mises Stress) Results

Fig. 11 displays the contour plot of the Von Mises stress result of samples for the all specimens that are prepared in this study. Fig. 12 illustrates the relationship between the results Von Mises stress and the type of samples. It can be noticed that the values of stress increased with the increased weight fraction of both types of particles for all groups. It can also be seen that the values of Von Mises stress for samples containing HA are higher than those of samples containing TiO_2.

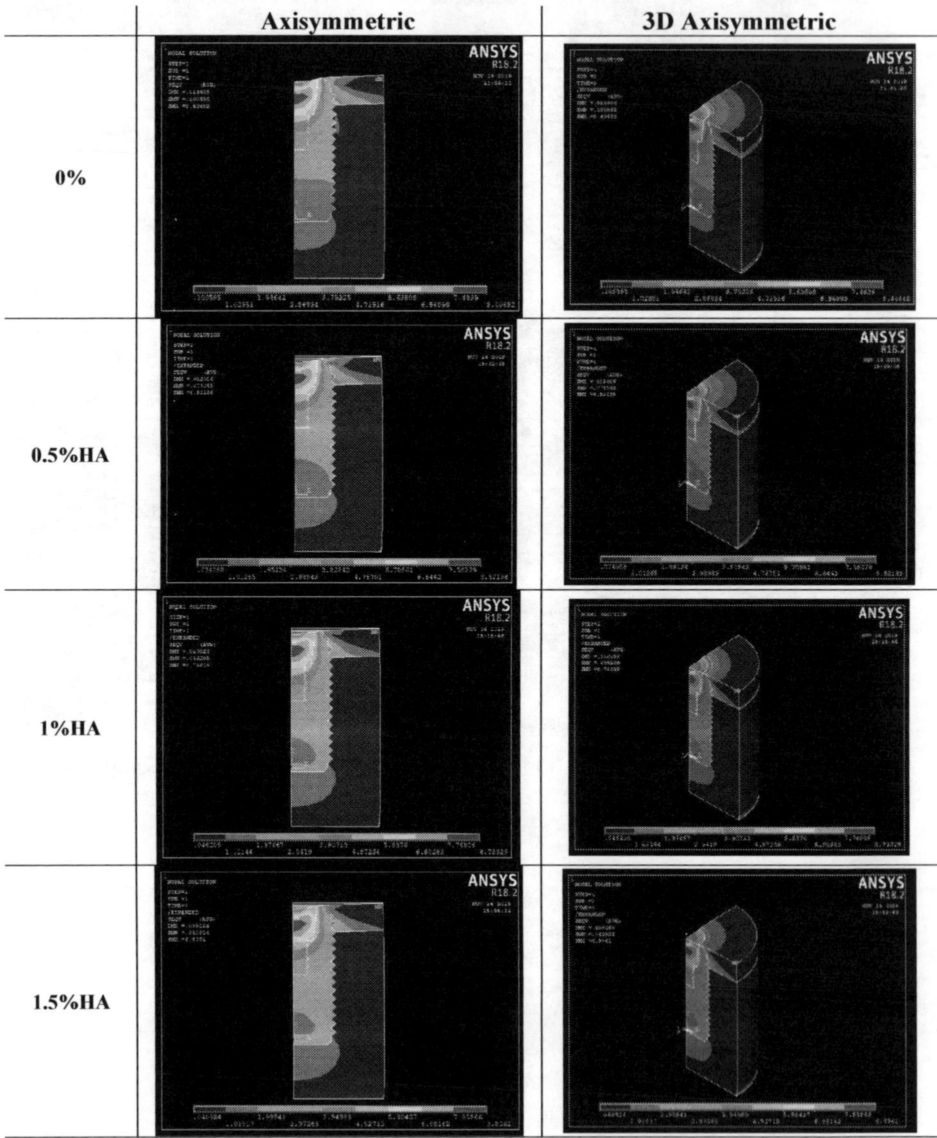

	Axisymmetric	3D Axisymmetric
0%		
0.5%HA		
1%HA		
1.5%HA		

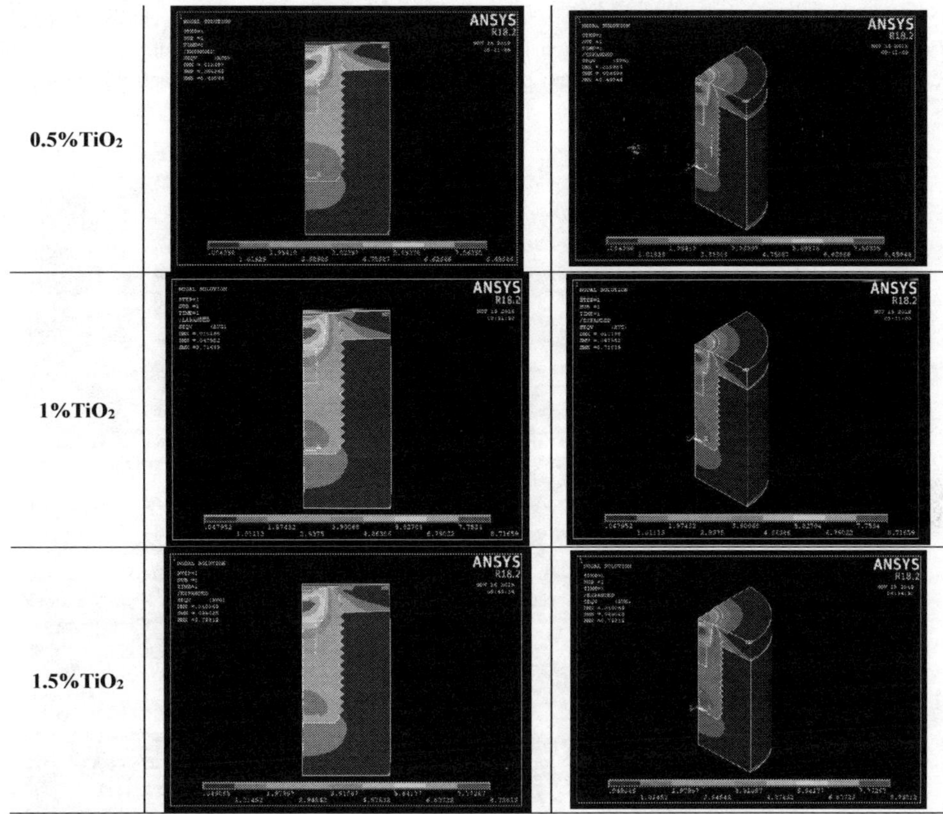

Fig.11 Maximum Von Mises stress result from the finite element analysis for the polymer composites

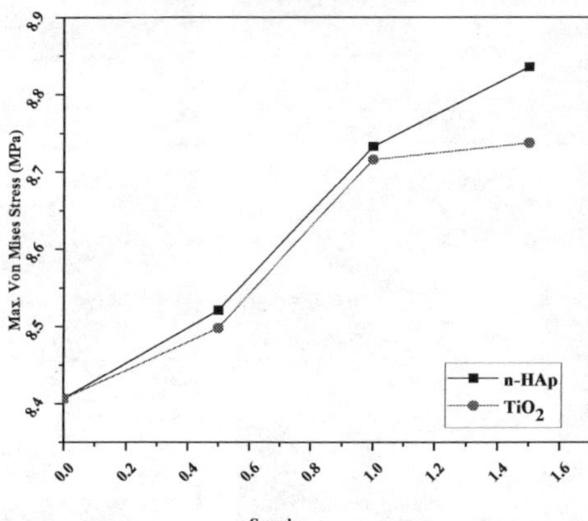

Fig. 12 Maximum Von Mises Stress result for the polymer composites

4.3.3 Equivalent Elastic Strain Results

Fig.13 manifests the contours plots of elastic strain of samples for all specimens that are prepared in this study. Fig.14 reveals the relation between the kind of samples and the elastic strain of the specimens. It can be noticed that the values of the elastic strain decreased with increasing the weight fraction of both types of particles for all groups. It can also be seen that the values of the elastic strain for samples containing HA are lower than those of the elastic strain for samples containing TiO_2. As well, it can also be noted from this figure that the highest value of the elastic strain occurred in the pure PEEK specimen is equal to (2.324×10^{-3}). But, the lowest value of the elastic strain that can be seen in the polymer specimens containing 1.5% HAp particles added to the PEEK matrix is equal to (1.269×10^{-3}).

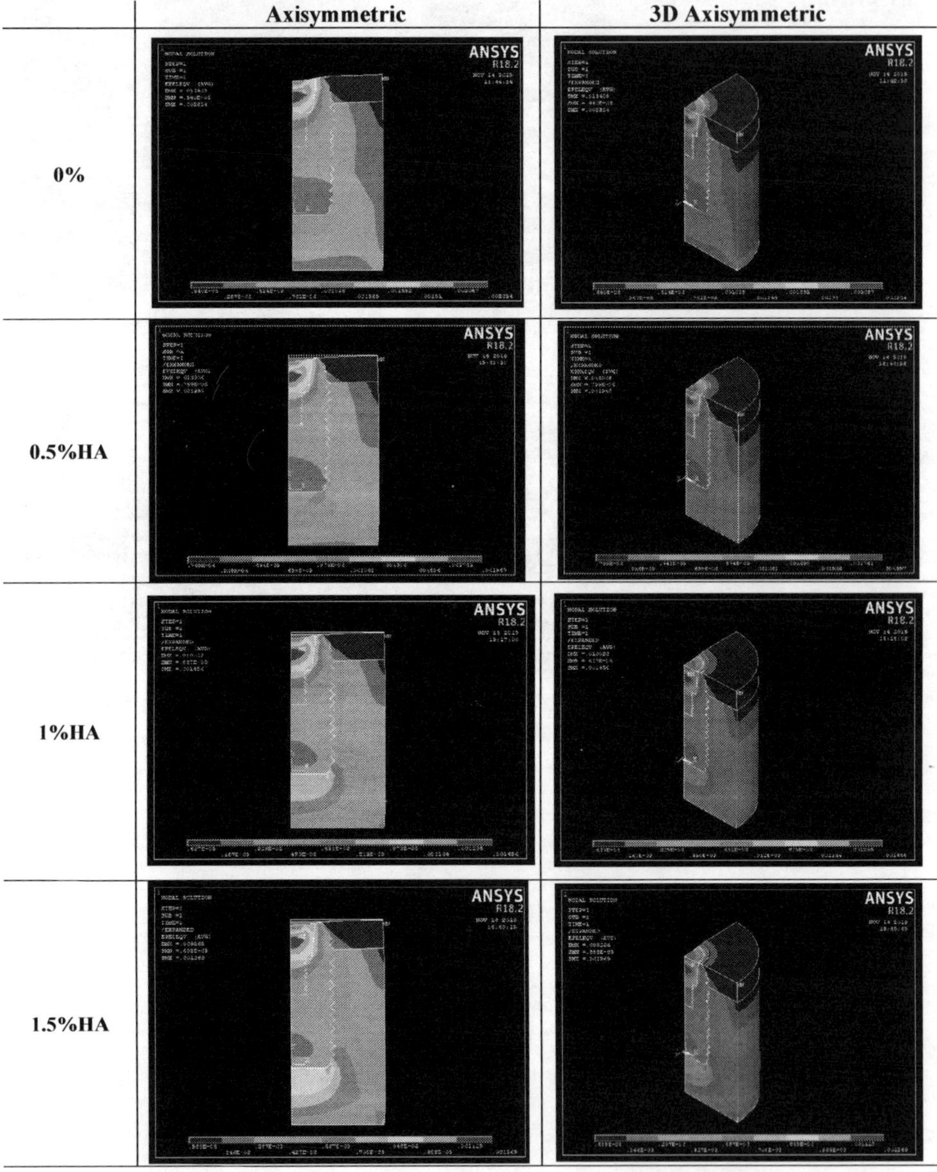

0.5%TiO_2

1%TiO_2

1.5%TiO_2

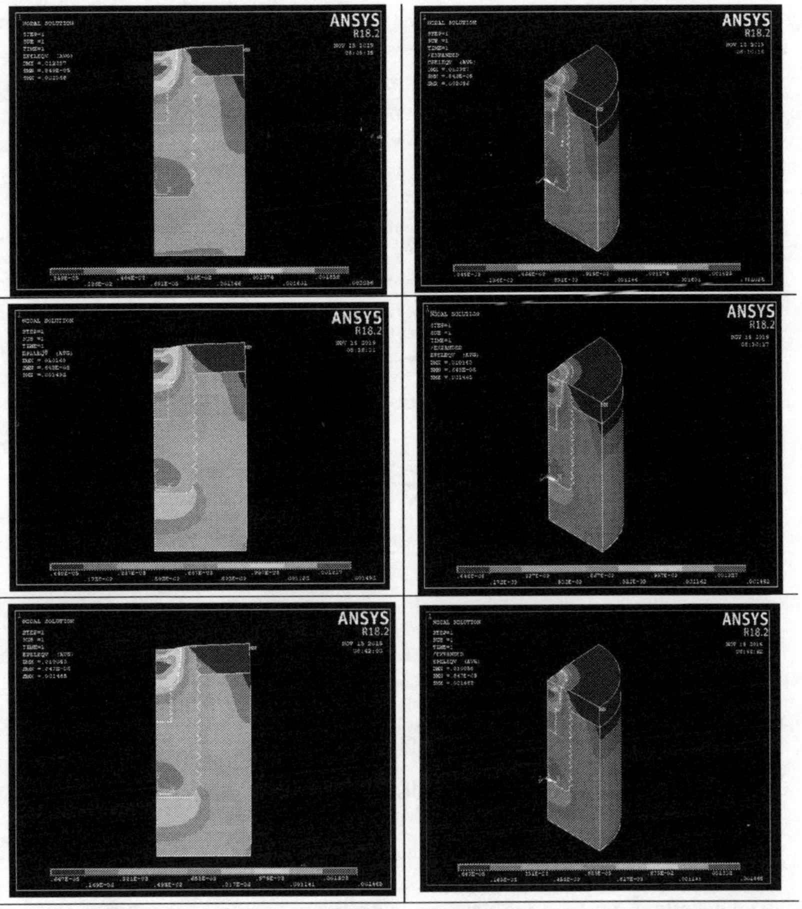

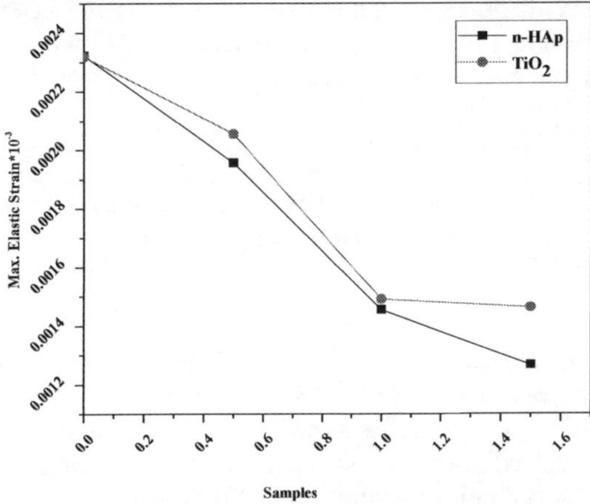

Fig.13 Maximum elastic strain results from the finite element analysis for the polymer composites

Fig.14 Maximum elastic strain results for the polymer composites

4.3.3 Theoretical Safety Factor Results

The safety factor of the specimens is obtained according to the Von Mises (Distortion Energy) failure theory. Fig. 15 elucidated the relationship between the type of samples and the theoretical safety factor of the specimens. From these figures, it can be observed that all specimens are in safe side, because the safety factor values are more than one indicating that the failure will not occur before the life of design is attained [42]. Furthermore, it can be noticed that the highest safety factor value occurred in polymer samples in the specimens containing 1.5% HAp is equal to (8.992). But, the lowest value of the safety factor that can be seen in the neat PEEK specimen is equal to (7.955).

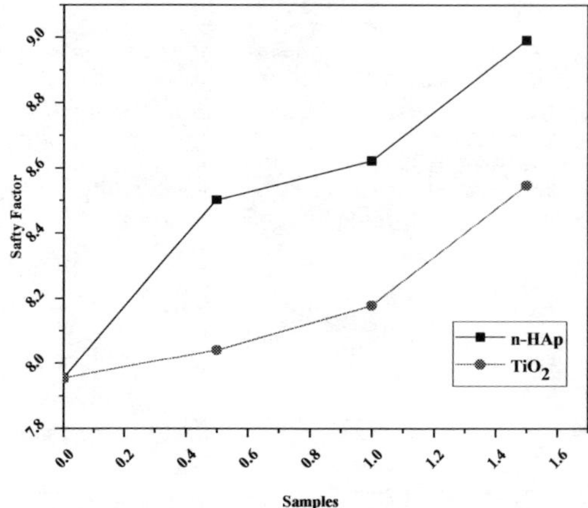

Fig.15 Theoretical safety factor results for the polymer composites

5. Conclusions

In this study, it was found that:
1. The thermal conductivity for the all samples decreased with the increased weight fraction of reinforcement.
2. The thermal analysis results explained the of (T_g and X_c%) and decrease in the value of (T_m and T_c) when adding nano particles.
3. The results ANOVA analysis are significant difference between the groups was a non-significant difference for the all samples ($P \geq 0.05$).
4. From the numerical results, the maximum Von Mises stress and safety factor were obtained for the samples containing 1.5% HAp, while the least was for the neat PEEK. Whereas, the total deformation and the maximum elastic strain have an opposite behavior.

Acknowledgements

The authors gratefully acknowledge the University of Technology-Iraq and Iran Polymer and Petrochemical Institute for their support and performing this work.

References

[1] M. Geetha, A.K. Singh, R. Asokamani, A.K. Gogia, Ti based biomaterials, the ultimate choice for orthopaedic implants - a review, Prog. Mater. Sci., 54 (2009) 397–425.

[2] E. Gibon, D.F. Amanatullah, F. Loi, J. Pajarinen, A. Nabeshima, Z.Y. Yao, M. Hamadouche, S.B. Goodman, The biological response to orthopaedic implants for joint replacement: part I: metals, J. Biomed. Mater. Res. B Appl. Biomater., 105 (2017) 2162–2173.

[3] M. Niinomi, M. Nakai, J. Hieda, Development of new metallic alloys for biomedical applications, Acta Biomater., 8 (2012) 3888–3903.

[4] J.W. Lee, H.B. Wen, P. Gubbi, G.E. Romanos, New bone formation and trabecular bone microarchitecture of highly porous tantalum compared to titanium implant threads: a pilot canine study, Clin. Oral Implants Res., 29 (2018) 164–174.

[5] A.J. Rahyussalim, A.F. Marsetio, I. Saleh, T. Kurniawati, Y. Whulanza, The needs of current implant technology in orthopaedic prosthesis biomaterials application to reduce prosthesis failure rate, J. Nanomater., 9 (2016), https://doi.org/10.1155/2016/5386924.

[6] G. Ryan, A. Pandit, D.P. Apatsidis, Fabrication methods of porous metals for use in orthopaedic applications, Biomaterials, 27 (2006) 2651–2670.

[7] Sagomonyants KB, Jarman-Smith ML, Devine JN, Aronow MS, Gronowicz GA, The in vitro response of human osteoblasts to polyetheretherketone (PEEK) substrates compared to commercially pure titanium, Biomaterials ,2008;29:1563–72.

[8] J. M. Toth, M. Wang, B. T. Estes, J. L. Scifert, H.B. Semi III and A. S. turner, Polyetheretherketone as a biomaterial for spinal applications, Biomaterials, 27(2006) 324-334.

[9] C. Dꞌıaz and G. Fuentes, Tribological Studies Comparison between UHMPE and PEEK for prosthesis application, Surface and Coatings Technology, 325(2017)656-660.

[10] J. Kolmas, S. Krukowski, A. Laskus and M. Jurkitewicz, Synthetic hydroxyapatite in pharmaceutical applications, Ceram. Int. ,42 (2016) 2472–2487.

[11] M.S. Abu Bakar, P. Cheang and K.A. Khor, Thermal processing of hydroxyapatite reinforced polyetheretherketone composites, Journal of Materials Processing Technology, 89-90 (1999) 462-466.

[12] S. Yousefali, A. Reyhani, S. Z. Mortazavi, N. Yousefali and A.Rajabpour, UV-blue spectral down-shifting of titanium dioxide nano-structures doped with nitrogen on the glass substrate to study its anti-bacterial properties on the E. Coli bacteria, Surfaces and Interfaces 13 (2018) 11–21.

[13] S. Shi, D. Shen, T. Xu and Y. Zhang, Thermal, optical, interfacial and mechanical properties of titanium dioxide/shape memory polyurethane nanocomposites, 164 (2018)17-23.

[14] K. Bataineh and M. Al Janaideh, Effect of different biocompatible implant materials on the mechanical stability of dental implants under excessive oblique load, Clinical Implant Dental and Related Research, 21(2019) 1206-1217.

[15] A. D. Schwitalla, M. Abou-Emara, T. Spintig, J. Lackmann and W. d. Muller, Finite Element Analysis of the Biomechanical Effects of PEEK Dental Implants on the Peri-implant Bone, Journal of Biomechanics, 48(2015)1-7.

[16] S. M. Tang, P. Cheang, M. S. AbuBakar, K. A. Khor, and K. Liao, Tension-tension fatigue behavior of hydroxyapatite reinforced polyetheretherketone composites, International Journal of Fatigue, 26(2004)49–57.

[17] K. L. Wong, C. T. Wong, W. C. Liu, H B Pan, M. K .Fong, W. M. Lam, W. L. Cheung, W. M. Tang, K. Y. Chiu, K. D. K. Luk, W. W. Lu, Mechanical properties and in vitro response of strontium-containing hydroxyapatite/polyetheretherketone composites, Biomaterials, 30(2009)3810–3817.

[18] I.Y. Kim, A. Sugino, K. Kikuta, C. Ohtsuki and S.B. Cho, Bioactive composites consisting of PEEK and calcium silicate powders, J. Biomater. Appl., 24(2009)105–18.

[19]L.Wang,S.He,X.Wu,S.Liang,Z.Mu,J.Wei,Polyetheretherketone/nanofluorohydroxyapatite composite with antimicrobial activity and osseointegration properties, Biomaterials, 35(2014)6758-6775.

[20] T.P. Oliveira, S. N. Silva and J. A. Sousa, Flexural Fatigue Behavior of Plasma-Sprayed Hydroxyapatite-Coated Polyether-ether-ketone (PEEK) Injection Moldings Derived from Dynamic Mechanical Analysis, International Journal of Fatigue, 108 (2018)1-8.

[21] K. Masamoto, S. Fujibayashi, T. Yabutsuka, T. Hiruta, B. Otsuki, Y. Okuzu, K. Goto, T. Shimizu, Y. Shimizu, C. Ishizaki, K. Fukushima, T. Kawai, M. Hayashi, K. Morizane, T. Kawata, M. Imamura and S. Matsuda, In vivo and in vitro bioactivity of a "precursor of apatite" treatment on Polyetheretherketone, Acta Biomaterialia, 91 (2019) 48-59.

[22] X. Lv, X. Wang, S. Tang, D. Wang, L. Yang, A. He, T. Tang and J. Wei, Incorporation of Molybdenum Disulfide into Polyetheretherketone Creating Biocomposites with Improved Mechanical, Tribological Performances and Cytocompatibility for Artificial Joints Applications, Colloids and Surfaces B: Biointerfaces, 189 (2020)110819.

[23] A. A. Mohammed, E. S. Al-Hassani, J. K. Oleiwi, S.R. Ghaffarian, The Effect of Annealing on The Behavior of Polyetheretherketone Composites Compared to Pure Titanium, Materials Research Express, 6 (2019)125405.

[24] A. A. Mohammed, E. S. Al-Hassani, J. K. Oleiwi, The nanomechanical characterization and tensile test of polymer nanocomposites for bioimplants, The 19th int. conference on technologies and materials for renewable energy, environment and sustainability-TMRESS, Lebanon, 2019.

[25] Annual Book of ASTM Standard, Standard Test Method for Transition Temperatures of Polymers by Differential Scanning Calorimetry, D 3418-03(2003)1-6.

[26] ASM Hand book, Properties and Selection Nonferrous Alloy and Special Purpose Materials", 5th Edition, 2(1998)897– 9010.

[27] Q. Li Zeng, Y. and X. tang, The Application and Research Progresses of Nickel-Titanium Shape Memory Alloy in Reconstructive Surgery ', Australia's phys. Eng. Sci. Med., 33(2010)129–136.

[28] A. A. Mohammed, E. S. Al-Hassani and J. K. Oleiwi, The Nanomechanical Characterization and Tensile Test of Polymer nanocomposites for Bioimplants, The 19th int. conference on technologies and materials for renewable energy, environment and sustainability-TMRESS, Lebanon, 2019.

[29] J. T. Strong and C. E. Misch, Functional Surface Area: Thread-form Parameter Optimization for Implant Body Design, Compend Contin Educ Dent, 19(1998)4–9.

[30] J. Zmudzki, G. Chladek and J. Kasperski, The Influence of A complete Lower Denture Destabilization on The Pressure of The Mucous Membrane Foundation, Acta of Bioengineering and Biomechanics, 14(2012)67-73.

[31] J. K. Oleiw, E. S. Al-hassani and A. A. Mohammed, Tensile and Buckling Analysis of Polymeric Composite Columns, Basrah Journal for Engineering Sciences, 14(2014)176-188.

[32] G. Tao and Z. Xia, Biaxial Fatigue Behavior of an Epoxy Polymer with Mean Stress Effect, International Journal of Fatigue, 31(2009)678-685.

[33] S. Choi and J. Kim, Thermal Conductivity of Epoxy Composites with A Binary-Particle System of Aluminum Oxide and Aluminum Nitride Fillers, Composites: Part B Engineering, 51(2013)140–147.

[34] K. Yangchuan, Z. Yubin and W. Zhongwen, The Measurements of Crystallinity Degree of PEEK, Chinese Journal Of Materials Research, 10(1996)205-209.

[35] J. Chen, Q. Guo, Z. Zhao, X. Shao, X. Wang and C. Duan, Thermal, Crystalline, and Tribological Properties of PEEK/PEI/PES Plastics Alloys, Journal of Applied Polymer Science, 127(2013)2220-2226.

[36] J. N. Panda, J. Bijwe and R. K. Pandey, Optimization of The Amount of Short Glass Fibers for Superior Wear Performance of PAEK Composites, Composites Part A: Applied Science and Manufacturing, 116(2019)158-168.

[37] B. Li, D. Liu, G. Li, X. Yang, Multifold Interface and Multilevel Crack Propagation Mechanisms of Graphene Oxide/Polyurethane/Epoxy Membranes Interlaminar-Toughened Carbon Fiber-Reinforced Polymer Composites, Journal of Materials Science 53(2018)15939-15951.

[38] X. Hou, Y. Hu, X. Hu and D. Jiang, Poly (ether ether ketone) Composites Reinforced by Graphene Oxide and Silicon Dioxide Nanoparticles, High Performance Polymers, 30(2018)406-417.

[39] X. Wang, J. Jin and M. Song, An Investigation of The Mechanism of Graphene Toughening Epoxy, Carbon, 65(2013)324-333.

[40] X. Wang and M. Song, Toughening of Polymers by Graphene, Nanomaterials and Energy, 2(2013)265-278.

[41] M. He, X. Chen, Z. Guo, X. Qiu, Y. Yang, C. Su, N. Jiang, Y. Li, D. Sun and L. Zhang, Super Tough Graphene Oxide Reinforced Polyetheretherketone for Potential Hard Tissue Repair Applications, Composites Science and Technology, 174(2019)194-201.

[42] S. L. Evans and P. J. Gregson, Composite Technology in Load-Bearing Orthopedic Implants, Engineering Materials, University of Southampton, Biomaterials, 19(1998)1329–1342.

Nano Hybrids and Composites
ISSN: 2297-3370, Vol. 33, pp 35-46
© 2021 Trans Tech Publications Ltd, Switzerland

Submitted: 2021-06-19
Revised: 2021-08-20
Accepted: 2021-09-01
Online: 2021-10-11

Electronic, Vibrational, and Structural Study of Polysaccharide Agar-Agar Biopolymer

Ankita Pandey[1,a], Abhishek Kumar Gupta[2,b]*, Shivani Gupta[3,c], Sarvesh Kumar Gupta[4,d] and Rajesh Kumar Yadav[5, e]

[1,2,3,4]Nanoionics and Energy Storage Laboratory (NanoESL), Department of Physics and Material Science, Madan Mohan Malviya University of Technology, Gorakhpur, India- 273010

[5]Department of Chemistry and Environmental Science, Madan Mohan Malaviya University of Technology, Gorakhpur (U. P.), India- 273010

[a]ankitapandey87706@gmail.com, [b]akgupta.mmmut@gmail.com, [c]shivanigkpdec10@gmail.com, [d]skdephysicist001@gmail.com, [e]rd100977@gmail.com

Keywords- DFT, HOMO-LUMO, biopolymer, FTIR, ESP.

Abstract. Polysaccharide biopolymer Agar-Agar extracted from red algae is a natural and biodegradable polymer. It is a combination of agarose (a neutral and linear polymer, with repeated units of agarobiose) and a heterogeneous mixture of agaropectin (a charged sulfated polymer). In this study, a comparative study of structural vibrational and electrochemical properties of agar-agar biopolymer with two different methods HF (Hartree-Fock) and DFT (Density Functional Theory) using a basis set 631+G (d, p) is performed. The comparative structural study of agar-agar biopolymer by HF and DFT method has been carried out to calculate the stability of the molecule. The thermionic properties and Mulliken charge distribution are analysed to deliver a quantitative study of partial atomic charge distribution. The overall vibrational analysis of primal modes of the biopolymer has been studied using FTIR analysis. Based on highest occupied molecular orbital (HOMO) and the lowest unoccupied molecular orbital (LUMO) composition and energies, various chemical parameters of the biopolymer have been evaluated. The Physico-chemical properties of this polysaccharide show a strong correlation with its optimized structure. Agar-agar has its application in the electrochemical, biotechnological, and pharmaceutical fields, as a stabilizer and gelling material.

Introduction

Around 140 million tons of synthetic polymers are manufactured worldwide each year. But synthetic polymers are quite expensive and cause environmental pollution such as food imitation, POPs (persistent organic pollutants or toxins) secretion, landfill accumulation, and the problem of ocean trash [1]. This problem gave rise to a new polymer that consists of living cells, is of low – cost, having degradation and renewable property, and is also environmentally friendly, known as a biopolymer.

Biopolymer or polymeric biomolecules, as the name suggests, are of biological origins such as alginate, chitosan, and carrageenan and having monomeric units that are covalently bonded [2]. The generation of biopolymer within the cell is caused by complex metabolic processes. Biopolymers are shaped into many forms like disks, membranes, beads, nanoparticles, and so on.

According to monomeric units, biopolymers are of three types, polynucleotides (such as DNA, RNA), polypeptides(protein), and polysaccharides (i.e., Polymeric carbohydrates, red algae). Polysaccharides are a class of natural polymers. They are abundant, renewable, non-toxic, low-cost, and biodegradable [3-8].

Having specific physical, chemical, biological, biomechanical, and degradation properties [9], biopolymers (also referred to as degradable polymeric biomaterials) have been widely used in immobilization techniques to save money and time, in the food industry as packing material [10, 11] in cosmetic aspects, in the medical industry [12], in automotive. In the field of high energy density batteries for electric propulsion and fuel cells for electrical vehicles. As an electrolytic material, polymer electrolyte occupies a definite place thanks to their varied super advantages like processability, flexibility, power density, cyclability, raw materials handiness, low weight, safety, and price.

Agar, a Mixture of linear polysaccharide agarose and agaropectin biopolymer [9], is obtained from gelidiella, gelidium, and gracilaria (i.e., red algae or red seaweeds) [13]. Gelidiella is mostly found in India mainly in the tropical and subtropical waters. On the other hand, Spain, Portugal, and Morocco-like countries are the broad source of gelidium harvesting while gracilaria are found in the cold water of Canada and some species live in warm, tropical climate water in Indonesia [14].

The composition of Agar is comparable to starch, which consists of amylose and amylopectin [15]. The most constituent of Agar is agarose that exhibits a composition $[C_{12}H_{14}O_5(OH)_4]_n$. Agaropectin, the other constituent of Agar, is a more complicated polysaccharide that could be a lot of sophisticated polyose with sulfuric elements and uronic acid residues [16]. Agar is utilised as a gelling, stabilising, and encapsulating ingredient in biomedical, food, and electrochemical energy storage technology [3]. The hydrophobicity nature of Agar increases its gelling and stabilizing tendency. In electrochemistry, agar biopolymer is used as salt bridges and gel plugs. Polysaccharide biopolymer has brought wide applications in all phases of oil recovery, from good drilling to waste product treatment. The biopolymers are typically used as additives of fluids or plugging agents to correct the fluid properties, and this affects the performance and price of oil recovery [15].

Marc Lahaye has reported the chemical structure and Physico-chemical property of Agar [9]. The structure of Agar constituent, agarose has been reported by Choji Araki [16]. Effect of Agar-based biopolymer electrolyte composed with NH$_4$I has been studied by S. Selvalakshmi and reported high conductivity of 1.17×10^{-4} S cm^{-1} at 303 K with low activation energy equal to 0.43 eV of polymer electrolyte having 50mol% agar doped with 50mol% NH$_4$I [3]. S. Smitha has been reported that how using Agar biopolymer improved the shear strength of Sabarmati sand [17]. The novelty of this work is that this work on biopolymer using DFT is new and has not been done before. To the best of the author's knowledge, there has been neither DFT nor HF study are reported on agar biopolymer in the literature so far and this is the beauty and novelty of this work. In this paper, we have investigated the structural, vibrational, and electrical properties of Agar-Agar biopolymer using the B3LYP and HF methods. For the theoretical study of biopolymer molecules and to explore the relationship between structural and electronic properties of chemical material, Hartree-Fock and density functional theory calculations having vibrational frequency is an essential and helpful tool [18, 19]. Mulliken population, molecular electrostatic potential, dipole moment, polarizability, and enthalpy are also reported here. The theoretically analyzed FTIR has been compared with the experimental values as reported by S. Selvalakshmi [3].

Computational Details

The entire simulations have been performed at Hartree-Fock and Density functional theory using Gaussian 09 [20] windows program involving gradient geometry optimization in the gas phase model. The structure of Agar-Agar was used as input for Hartree-Fock and B3LYP method [21] for optimization with density functional theory using a basis set 6-31+G(d,p). To calculate vibrational frequencies at the DFT levels, the optimized structural parameters have been used which characterized all fixed points as minima [22]. Among all the standard functionals, Becke's non – local gradient corrected density functional theory with three parameters hybrid functional(B3LYP) developed by Lee-Yang-Parr has been used to calculate the molecular structure, vibrational frequencies, and electronic parameters [23]. The calculated harmonic vibrational frequencies are used to study the number of imaginary frequencies and zero-point vibrational energy of the material [5, 21].

Using the principle of statistical mechanics, thermodynamic properties like Gibb's free energy, entropy and enthalpy are also examined. Enthalpy is calculated by victimization the subsequent equation [21]

$$\Delta S = \frac{\Delta H - \Delta G}{T} \qquad (1)$$

Where ΔG is Gibbs free energy, ΔH is entropy, ΔS is enthalpy and T is the room temperature. During this sequence, the other two more properties like polarizability and dipole moment that is the first derivative of energy concerning an applied field and is the measure of the asymmetry in the molecular charge distribution, have been also studied to characterize piezoelectric material. The Mulliken population analysis helps to study the estimation of partial atomic charges [24].

To analyze electronic and chemical stabilities, we have calculated the HOMO-LUMO energies levels for the calculation of the bandgap (E_g), fermi level energy (E_{FL}), ionization potential (IP), electron affinity (EA), electronegativity(χ), chemical potential(μ), chemical hardness(η), chemical softness(S), electrophilicity index(ω). By Koopmans' theorem, these numerous physical and chemical parameters will be calculated from HOMO-LUMO energy values according to the equations given below: [24, 25]

$$\text{Band Energy, } E_g = E_{HOMO} - E_{LUMO} \qquad (2)$$

$$\text{Fermi energy level, } E_{FL} = \frac{E_{HOMO} - E_{LUMO}}{2} \qquad (3)$$

$$\text{Ionization potential, } IP = -E_{HOMO} \qquad (4)$$

$$\text{Electron affinity,} EA = -E_{LUMO} \qquad (5)$$

$$\text{Electronegativity, } \chi = \frac{IP+EA}{2} \qquad (6)$$

$$\text{Chemical potential, } \mu = -\frac{(IP+EA)}{2} \qquad (7)$$

$$\text{Chemical hardness, } \eta = \frac{IP-EA}{2} \qquad (8)$$

$$\text{Chemical softness, } S = \frac{1}{\eta} \qquad (9)$$

$$\text{Electrophilicity index, } \omega = \frac{\mu^2}{2\eta} \qquad (10)$$

Result and Discussion

Structural analysis

The optimized structure of the Agar-Agar biopolymer is shown in Figure 1. The whole optimization is performed using B3LYP and Hartree-Fock (HF) method with a basis set 6-31+G (d, p). The optimized structural parameters of studied Agar like bond length (in Å), bond angle (in degree), and dihedral angle (in degree) obtained by B3LYP, and HF have been listed in the table for investigating the stability of the material. The bond lengths between C-C, C-O, C-H, O-H are approx. 1.532, 1.431, 1.092, 0.968 Å respectively obtained the same by HF and B3LYP both. The bond angle between C-C-C, C-O-C, 0-C-C, O-C-O, C-O-H, H-C-H, O-C-H are approx. 110.35°, 114.26°, 114.23°, 107.81°, 107.17°, 104.3°, 108.2° respectively, while the dihedral angle between C-C-C-O, C-O-C-O, C-C-O-C, C-O-C-C, O-C-C-O, H-C-H, C-C-C-C are -172.63°, -177.48°, -170.98°, -46.76°, -81.53°, - 38.09°, -58.97°.

Thus, the whole structural analysis can be concluded from the reported data in Table 1 that there is no significant difference between the geometrical parameters calculated by B3LYP and HF.

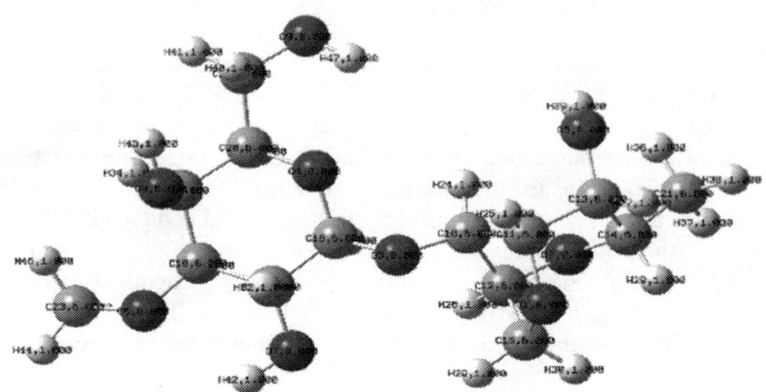

Fig. 1. The molecular structure or geometry optimization of AGAR -AGAR biopolymer.

Table 1. Geometrical parameters of AGAR-AGAR biopolymer.

Structural parameters		Method	
		DFT	*HF*
Bond length (Å)	O1-C11	1.437	1.437
	O1-C15	1.448	1.448
	C15-C12	1.532	1.532
	O2-C14	1.443	1.443
	O2-C12	1.431	1.431
	C13-C11	1.546	1.546
	O9-C22	1.418	1.418
	C12-C10	1.540	1.540
	C18-C19	1.538	1.538
	O7-C17	1.421	1.421
	O6-C18	1.419	1.419
	O6-C23	1.420	1.420
	O4-C20	1.428	1.428
	O4-C16	1.432	1.432
	C19-C20	1.543	1.543
	C18-C17	1.535	1.535
	O6-C18	1.419	1.419
	O8-C19	1.421	1.421
	O5-C13	1.431	1.431
	C11-C13	1.546	1.546
	C11-C10	1.525	1.525
	O3-C16	1.387	1.387
	O3-C10	1.429	1.429
	C14-C21	1.520	1.520
	C13-C14	1.546	1.546
	C16-C17	1.530	1.530
	C10-H24	1.092	1.092
	O7-H42	0.968	0.968
Bond angle (degree)	C23-O6-C18	115.42	115.42
	C19-C18-O6	115.05	115.05
	C18-C17-C16	110.35	110.35
	C20-O4-C16	114.26	114.26
	O8-C19-C20	114.23	114.23

	C20-C22-O9	111.73	111.73
	C17-C18-C19	111.50	111.50
	O3-C16-O4	107.81	107.81
	C21-C14-O2	107.03	107.03
	O2-C13-C14	112.71	112.71
	C14-C13-O5	110.18	110.18
	C14-O2-C12	115.16	115.16
	O2-C12-C10	107.63	107.63
	O2-C12-C15	113.95	113.95
	C18-C17-O7	110.11	110.11
	C16-C17-O7	107.63	107.63
	C10-O3-C16	115.32	115.32
	O3-C10-C12	112.39	112.39
	C10-C12-C15	100.92	100.92
	C12-C15-O1	105.24	105.24
	C11-O1-C15	108.22	108.22
	C21-C14-C13	113.86	113.86
	C13-C11-C10	108.48	108.48
	O5-C13-C11	111.04	111.04
	C14-C13-O5	110.18	110.18
Dihedral angle (degree)	O8-C19-C18-O6	51.55	51.55
	C23-O6-C18-C19	67.58	67.58
	C23-O6-C18-C17	-168.91	-168.91
	O6-C18-C17-O7	61.65	61.65
	O6-C18-C17-C16	-179.67	-179.67
	O7-C17-C16-O3	-68.22	-68.22
	C17-C16-O3-C10	157.59	157.59
	C17-C1-O4-C20	-58.68	-58.68
	C17-C18-C19-C20	54.53	54.53
	O8-C19-C20-C22	-58.05	-58.05
	C19-C18-C17-O7	-172.63	-172.63
	C20-O4-C16-O3	-177.48	-177.48
	C21-C14-C13-O5	40.63	40.63
	C21-C14-O2-C12	-170.98	-170.98
	C14-O2-C12-C15	-46.76	-46.76
	O2-C14-C13-O5	-81.53	-81.53
	C12-C10-O3-C16	-91.11	-91.11
	C10-O3-C16-O4	-82.72	-82.72
	O5-C13-C11-O1	176.85	176.85
	C11-O1-C12-O2	75.24	75.24
	C15-C12-C10-C11	44.97	44.97
	C13-C11-C10-O3	-169.82	-169.82
	C14-C13-C11-C10	-58.97	-58.97
	O3-C10-C12-O2	170.78	170.78
	H26-C12-C15-H29	-38.09	-38.09

Thermal property

Table 2 contains the thermionic property of agar-agar biopolymer calculated by HF and B3LYP method. The total energy, dipole moment, and polarizability are assessed to the degree of exactness. Dipole moment and polarizability are two significant properties to describe a piezoelectric material. The dipole moment is the principal subordinate of energy concerning an applied electric field whereas polarizability is quantified as the second derivative of the energy of the molecule with relevance to an electrical field. Gibbs free energy(ΔG), entropy(ΔS), and enthalpy(ΔH) which is obtained by using Equation 1 have been calculated and reported in Table 2. Gibbs free energy is the estimation of the maximum reversible work in any reversible system at a constant temperature and pressing factor i.e., pressure [21, 24]. Here the positive Gibbs free energy value indicates the non-spontaneous process. The calculated value of ΔG by HF and B3LYP is approx. 237.826 and 185.74 (kcal/mol) respectively.

Table 2. Thermionic parameters of AGAR-AGAR biopolymer.

Properties	HF	B3LYP
Electronic energy(kcal/mol)	-764233.91	-768667.89
Dipole moment (Debye)	3.538	3.108
Polarizability (au.)	175.82	195.23
Zero-point energy(kcal/mol)	266.25	220.88
Enthalpy (kcal/mol)	278.9	262.7
Gibb's free energy(kcal/mol)	237.826	185.74
Entropy (Cal/mol-ke)	137.977	174.243

Mulliken population analysis

Mulliken atomic charges of studied biopolymer structure have been carried out by using HF and B3LYP method in the gaseous phase. The calculated value of Mulliken atomic charge distribution is reported in Table 3. All the oxygen atoms of the optimized agar molecule are gained a negative charge value due to their more electronegative nature than the other two atoms and act as a donor [26]. Carbon exhibits positive and negative Mulliken atomic charge values. The negative charge value of carbon atoms is found where it attaches with hydrogen only since hydrogen is less electronegative than carbon while the positive charge values are displayed where it attaches with an oxygen atom as O is more electronegative than carbon. All hydrogen atoms act as accepters, exhibit a net positive charge. The maximum charge value of hydrogen has been observed at H42 i.e., 0.3927 due to the presence of an electronegative atom (O7).

Table 3. Mulliken charge analysis of AGAR-AGAR biopolymer.

Atom	Charge	Atom	Charge	Atom	Charge	Atom	Charge	Atom	Charge
O1	-0.5543	C11	-0.0883	C21	-0.5198	H31	0.0979	H41	0.1293
O2	-0.4526	C12	0.2805	C22	0.0646	H32	0.1609	H42	0.3927
O3	-0.3652	C13	0.0912	C23	0.0272	H33	0.1144	H43	0.3655
O4	-0.5033	C14	0.2444	C24	0.1671	H34	0.1053	H44	0.1333
O5	-0.6147	C15	0.0311	C25	0.1372	H35	0.1221	H45	0.1034
O6	-0.4423	C16	-0.0675	C26	0.1404	H36	0.1593	H46	0.1164
O7	-0.6285	C17	0.5014	H27	0.1240	H37	0.1402	H47	0.3822
O8	-0.6449	C18	-0.2026	H28	0.1170	H38	0.1204	-	-
O9	-0.5886	C19	0.4255	H29	0.1448	H39	0.3694	-	-
C10	-0.1161	C20	0.0449	H30	0.1266	H40	0.1081	-	-

Vibrational study

To investigate the structure of Agar-Agar, the FTIR analysis has broad applicability [3]. In this vibrational study, the FTIR spectra of Agar-Agar at the HF and B3LYP level with 6-31+G(d,p) basis set have been calculated by using optimized structural parameters. The observed FTIR spectra of Agar-Agar by methods HF and DFT are shown in Figure 2. The total 42 no. of atoms consist of Agar molecule, and it was expected to have 135 normal modes of vibrations. Table 4 represents the observed and calculated frequencies and their corresponding vibrational assignments of Agar-Agar biopolymer. The FTIR spectra by B3LYP and HF is shown in Figures 1 and 2. The 3879- 3800 cm^{-1} region is assigned to O-H stretching while O-H in- plane-bending lies in the region 1155 cm^{-1} and 1078 cm^{-1} by HF and DFT method, respectively. The C-C stretching is found in the region 1600-

1423 cm^{-1} by HF and by DFT it is found in the region 1435-1298 cm^{-1} whereas the C-C bending lies in the region 1380 cm^{-1} and 1239 cm^{-1} by HF and DFT respectively [23]. The FTIR band observed at 3879 is assigned to O-H stretching. The characteristic peak of Agar-Agar by HF method is found at frequency 1212 cm^{-1} which is assigned to C-O-C stretching while by B3LYP method it is found at frequency 1146 cm^{-1} and this is assigned to C-O-C bending. The vibrational frequency obtained by using HF and DFT method is nearly the same as experimentally reported by S. Selvalakshmi.[3]

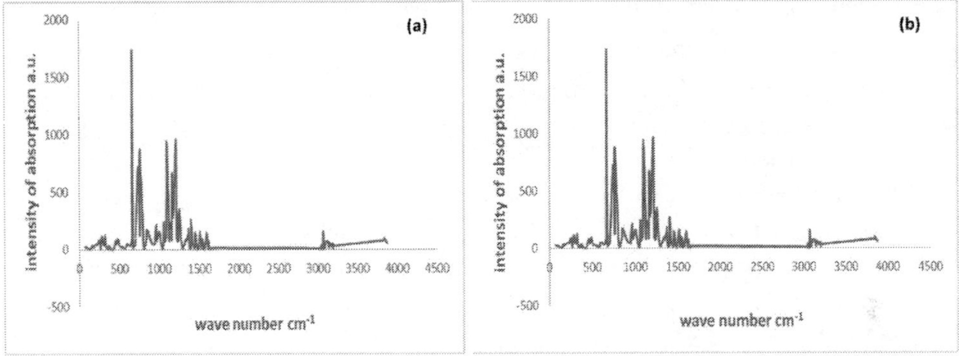

Fig. 2. Observed FTIR spectrum of Agar-agar (a) by HF method (b) by DFT method.

Table 4. The calculated frequency with observed assignments by HF and DFT methods.

Observed Wavenumber (cm^{-1}) (HF)	Observed Wavenumber (cm^{-1}) (DFT)	Experimentally obtained	Assignment
3879-3841	3837-3800	3404	O-H stretching
3204-3124	3151-3124		CH$_3$ stretching
3071	3092	2924	CH$_2$ stretching
1600-1423	1435-1298		C-C stretching
1434	1290	1392	CH$_2$ bending
1380	1239		C-C bending
1366	1168		CH bending
1155	1078		O-H bending
1058	1032	1039	CH$_2$ scissoring

Frontier molecular orbital analysis

The HOMO-LUMO together are referred to as Frontier Molecular orbital (FMO$_s$). FMO plays a very important role in electrical properties, optical properties, luminescence, chemical reaction, UV – VIS, quantum chemistry, and pharmaceutical studies and additionally, it's accustomed study the biological mechanism of molecules. The chemical reactivity and therefore the kinetic stability of the molecule may be characterized with the assistance of the Frontier orbital gap. Besides this, the foremost reactive position of the studied molecule may be foretold by FMO$_S$ [27]. The value of highest occupied and lowest unoccupied orbital energy defines the electro denoting and receiving ability of a molecule. The position and distribution of HOMO and LUMO in Agar-Agar are shown in Figure 3. The obtained value of HOMO and LUMO orbital energy by the two different techniques HF and DFT are approx. (-10.75 and 1.626) and (-6.725 and -0.449) eV respectively. The bandgap energy or FMO$_S$ energy gap ($\Delta E_{HOMO-LUMO}$) of Agar-Agar biopolymer is found to be -12.38eV by HF and -6.276eV by DFT. In contrast to HF, delocalization error is caused by the approximate exchange in DFT, which does not cancel the self-interaction in J (Coulomb integral). When J is positive and K (electron affinity) is negative, the HOMO becomes unstable in DFT due to the excess positive J term [28, 29]. The chemical reactivity parameters of studied material like chemical potential, electronegativity, hardness, softness, and electrophilicity index have been calculated by using HOMO and LUMO orbital energy values reported in Table 5. The width of the HOMO-LUMO gap indicates chemical stability (large gap) or reactivity (small gap). The latter could be the result of a low-lying LUMO or a high-lying HOMO, which could favor a nucleophilic or electrophilic assault on a particular molecule. Also, since the atoms with the highest orbital coefficients in the HOMO are vulnerable to transfer electron density, whereas those in the LUMO suggest acceptor activity, the composition of the two frontier levels may predict reaction regioselectivity. In other words, the structure of the HOMO and LUMO means that the reactivity is regulated orbitally [30-33].

Table 5. Chemical parameters of AGAR-AGAR biopolymer obtained from HOMO-LUMO energy.

Parameter	Hartree-Fock	DFT
Band gap energy (E_g) (eV)	-12.38	-6.276
Fermi level energy (E_{FL}) (eV)	-6.19	-3.138
Ionization energy (IP) (eV)	10.754	6.725
Electron affinity (EA) (eV)	-1.626	0.449
Electronegativity (eV)	4.564	3.587
Chemical potential (eV)	-4.564	-3.587
Chemical hardness (eV)	6.19	3.138
Chemical softness (eV)	0.162	0.319
Electrophilicity index	1.683	2.0501

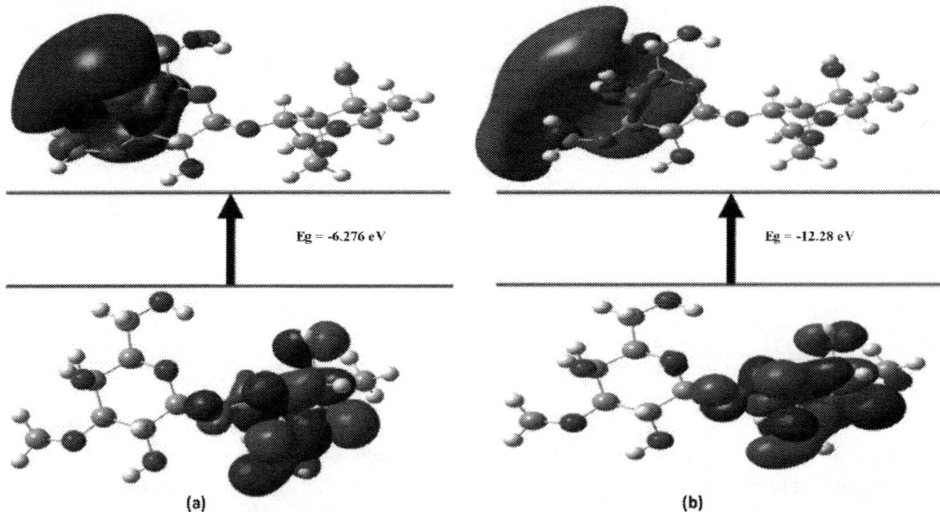

Fig. 3. Bandgap energy and HOMO-LUMO position of AGAR-AGAR biopolymer (a) by B3LYP/6-31+G(d,p), (b) by HF/6-31+(d,p).

Molecular electrostatic potential analysis

To study molecular modeling and to grasp molecular interaction, the molecular electrostatic potential is an important and widely used tool. The phenomenon that how different geometry can interact and the relative reactivity position for nucleophilic and electrophilic attacks are predicted by using the molecular electrostatic potential (MEP) [27]. Figure 4 shows the molecular electrostatic potential mapped surface as well as a mapped surface contour too. MEP is the study of the potential energy of a proton at a specific location near a molecule and the relative polarity of a molecule can be understood by the ESP analysis. The color coding refers to the positive, negative, and neutral electrostatic potential region. In other words, it illustrates the charge distribution over the region. The more electron-rich region is illustrated by blue color while the poor electron region is illustrated here by blue color. The attraction of the proton by the concentrated electron density within the molecules corresponds to the negative electrostatic potential that is colored in red shades wherever as the repulsion of the proton by the atomic nuclei in the regions where low electron density exists and the atomic charge is deficiently shielded corresponded to the positive electrostatic potential which is colored in the blue shades [26]. The blue color is mostly concentred on the oxygen atom that means the maximum charge density is obtained around oxygen. This shows the oxygen atom is most electronegative and carbon and oxygen atoms are attracted towards the oxygen atom. The greenish-yellow part denotes the neutral or negligible charge density. On the other hand, we have observed that the surface contour or waves are overlapped. This concludes that the bonding nature is partially ionic and partially covalent.

The MEP surface of the Agar-Agar biopolymer reported in Figure 4 has been determined by the B3LYP and HF calculation using the optimized structure with a 6-31G+(d,p) basis set. By HF and DFT, the color code of the titled material is found in the range of -6.039e^{-2} to +6.039e^{-2} au. After comparing the ESP surface mapped by the two different methods HF and B3LYP we have found that there is a very minor or negligible difference between the mapped surface formed by HF and DFTs.

-6.915e-2 a.u.   +6.915e-2 a.u.

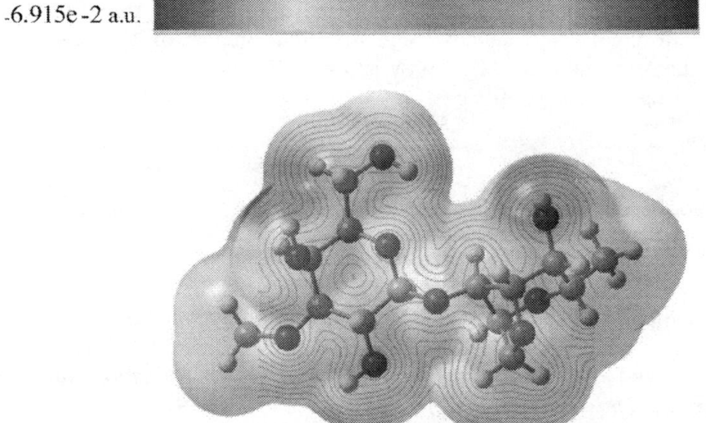

Fig. 4. The mapped MESP surface of agar.

Conclusion

Because oil and gas will continue to be the primary source of energy for society's progress, scientists and engineers have long sought to build an effective and ecologically friendly petroleum recovery process. To solve this, using biopolymer as a fluid addition can help increase oil recovery while also lowering drilling, hydrofracking, and wastewater treatment costs. Biopolymers are more efficient and cost less than manufactured polymers. The geological conditions and operating procedures, on the other hand, have a significant impact on biopolymer performance. Because these biomacromolecules' functions are dependent on their molecular conformations, defining and hence altering their structure is a viable way to improve their thickening, crosslinking, and adsorption effects. In this study, the structural, vibrational, and electrochemical properties of Agar-Agar biopolymer have been studied using density functional theory at the HF/6-31+G(d,p) and B3LYP/6-31+G(d,p) and compared each other. Crystal data yielded bond lengths and angles that are extremely close to experimental values. According to our structural analysis, we have not found a single difference between the values bond length, bond angle, and dihedral angle calculated by the two different methods HF and B3LYP. The thermodynamic functions of studied material like polarizability, dipole moment, and Gibbs free energy have been also reported here. Mulliken population analysis has been applied to quantify the charge distribution inside the optimized system and we have found that all oxygen atoms have negative charge values, carbon has negative and positive charge values while hydrogens exhibit only positive charge values. In the analysis of vibrational properties, the assignments obtained by the HF and DFT methods show agreement with the experimental data reported in the literature. In MEP analysis, it is found that the oxygen atom is surrounded by high charge density. The obtained HOMO-LUMO energies helped to study the chemical reactivity properties. The electrostatic potential of the molecule under investigation indicated attackable areas for electrophilic and nucleophilic chemicals. The addressed molecule is a soft molecule with a high chemical reactivity. Overall, the studied properties of AGAR-AGAR biopolymer indicate several applications in various fields of innovation like automobile, biopolymer, biomedical, petroleum recovery, etc. It could be used as plugging agents to correct fluid properties. Overall, we aim to gain a deeper understanding of the physical and chemical properties of biopolymers in the near future and then deploy biopolymers that are appropriate for specific geographical and energy storage conditions.

Acknowledgments

The authors would like to acknowledge the support provided by TEQIP-III, Madan Mohan Malaviya University of Technology, Gorakhpur, India in the form of a research initiation grant (No.-RIG/975/2018). Authors would like to acknowledge the kind support of Dr. Amarjeet Yadav, Babasaheb Bhimrao Ambedkar University, Lucknow is also acknowledged for helping in carrying out the simulation work using Gaussian software.

Conflict of Interest

The authors declare no conflict of interest.

References

[1]　A. Kawalkar. A Comprehensive Review on Osteoporosis. *J. Trauma*, vol. 10, no. 1, pp. 3–12 (2015).

[2]　M. E. Hassan, J. Bai, and D. Q. Dou. Biopolymers; Definition, classification and applications. *Egyptian Journal of Chemistry*, vol. 62, no. 9. pp. 1725–1737 (2019).

[3]　S. Selvalakshmi, T. Mathavan, S. Selvasekarapandian, and M. Premalatha. Study on NH4I composition effect in agar–agar-based biopolymer electrolyte. *Ionics (Kiel).*, vol. 23, no. 10, pp. 2791–2797 (2017).

[4]　J. W. Rhim and P. Kanmani. Synthesis and characterization of biopolymer agar mediated gold nanoparticles. *Mater. Lett.*, vol. 141, pp. 114–117 (2015).

[5]　M. S. H. Al-Furjan, M. Habibi, J. Ni, D. won Jung, and A. Tounsi. Frequency simulation of viscoelastic multi-phase reinforced fully symmetric systems. *Eng. Comput.*, no. 0123456789 (2020).

[6]　S. A. & A. T. Ehsan Arshid, Mohammad Khorasani, Zeinab Soleimani-Javid. Porosity-dependent vibration analysis of FG microplates embedded by polymeric nanocomposite patches considering hygrothermal effect via an innovative plate theory. *Eng. Comput.*, (2021).

[7]　X. Huang, H. Hao, K. Oslub, M. Habibi, and A. Tounsi. Dynamic stability/instability simulation of the rotary size-dependent functionally graded microsystem. *Eng. Comput.*, (2021).

[8]　M. S. H. Al-Furjan, A. hatami, M. Habibi, L. Shan, and A. Tounsi. On the vibrations of the imperfect sandwich higher-order disk with a lactic core using generalize differential quadrature method. *Compos. Struct.*, vol. 257, p. 113150 (2021).

[9]　M. Lahaye and C. Rochas. Chemical structure and physico-chemical properties of agar. *Hydrobiologia*, vol. 221, no. 1, pp. 137–148 (1991).

[10]　T. Rajesh Kumar Dora, S. Ghosh, and R. Damodar. Synthesis and evaluation of physical properties of Agar biopolymer film coating-an alternative for food packaging industry. *Mater. Res. Express*, vol. 7, no. 9 (2020).

[11]   K.H.B. Rachid Zerrouki, Abdelkader Karas, Mohamed Zidour, Abdelmoumen Anis Bousahla, Abdelouahed Tounsi, Fouad Bourada, Abdeldjebbar Tounsi. Effect of nonlinear FG-CNT distribution on mechanical properties of functionally graded nano-composite beam. *Struct. Eng. Mech.*, vol. 78, no. 2, pp. 117–124 (2021).

[12]   J. Jang and P. Jia. A Review of the Application of Biopolymers on Geotechnical Engineering and the Strengthening Mechanisms between Typical Biopolymers and Soils. *Advances in Materials Science and Engineering*, vol. 2020. (2020).

[13]   M. Dosta, K. Jarolin, and P. Gurikov. Modelling of mechanical behavior of biopolymer alginate aerogels using the bonded-particle model. *Molecules*, vol. 24, no. 14, pp. 1–15 (2019).

[14]   C. Araki. Structure of the Agarose Constituent of Agar-agar. *Bull. Chem. Soc. Jpn.*, vol. 29, no. 4, pp. 543–544 (1956).

[15]   S. Xia, L. Zhang, A. Davletshin, Z. Li, J. You, and S. Tan. Application of polysaccharide biopolymer in petroleum recovery. *Polymers*, vol. 12, no. 9. pp. 1–36 (2020).

[16]   S. Smitha and A. Sachan. Use of agar biopolymer to improve the shear strength behavior of sabarmati sand. *Int. J. Geotech. Eng.*, vol. 10, no. 4, pp. 387–400 (2016).

[17]   R. Sarkar and T. K. Kundu. Density functional theory studies on PVDF/ionic liquid composite systems. *J. Chem. Sci.*, vol. 130, no. 8 (2018).

[18]   S. PAL and T. K. KUNDU. DFT-based inhibitor and promoter selection criteria for pentagonal dodecahedron methane hydrate cage. *J. Chem. Sci.*, vol. 125, no. 5, pp. 1259–1266 (2013).

[19]   S. Pal and T. K. Kundu. Stability Analysis and Frontier Orbital Study of Different Glycol and Water Complex. *ISRN Phys. Chem.*, vol. 2013, pp. 1–16 (2013).

[20]   "Gaussian 09 Revision(B.01)," *(Gaussian Inc. Wallingford CT)* (2010).

[21]   S. U. D. Shamim *et al.*, A DFT study on the geometrical structures, electronic, and spectroscopic properties of inverse sandwich monocyclic boron nanoclusters ConBm (n = 1.2; m = 6–8). *J. Mol. Model.*, vol. 26, no. 6, (2020).

[22]   D. Farmanzadeh, A. Soltanabadi, and S. Yeganegi. DFT study of the geometrical and electronic structures of geminal dicationic ionic liquids 1,3-bis[3-methylimidazolium-1-yl]hexane halides. *J. Chinese Chem. Soc.*, vol. 60, no. 5, pp. 551–558 (2013).

[23]   S. Prabhakaran and M. J. M. Jaffar. Vibrational analysis, ab initio hf and DFT studies of 2,4,6-Trimethyl phenol. *Indian J. Pure Appl. Phys.*, vol. 56, no. 2, pp. 119–127 (2018).

[24]   M. A. Mumit, T. K. Pal, M. A. Alam, M. A. A. A. A. Islam, S. Paul, and M. C. Sheikh. DFT studies on vibrational and electronic spectra, HOMO–LUMO, MEP, HOMA, NBO and molecular docking analysis of benzyl-3-N-(2,4,5-trimethoxyphenylmethylene) hydrazinecarbodithioate. *J. Mol. Struct.*, vol. 1220, p. 128715 (2020).

[25]   T. Abbaz, A. Bendjeddou, and D. Villemin. Molecular orbital studies (hardness , chemical potential , electro negativity and electrophilicity) of TTFs conjugated between. *Int. J. Adv. Sci. Eng. Technol.*, vol. 5, no. 2, pp. 5150–5161 (2018).

[26]   M. M. Sanagi *et al.*, Agarose- and alginate-based biopolymers for sample preparation: Excellent green extraction tools for this century. *Journal of Separation Science*, vol. 39, no. 6. pp. 1152–1159 (2016).

[27]   L. Altomare *et al.*, Biopolymer-based strategies in the design of smart medical devices and artificial organs. *International Journal of Artificial Organs*, vol. 41, no. 6. pp. 337–359, (2018).

[28]   Y. Huang, C. Rong, R. Zhang, and S. Liu. Evaluating frontier orbital energy and HOMO/LUMO gap with descriptors from density functional reactivity theory. *J. Mol. Model.*, vol. 23, no. 1 (2017).

[29]   L. Kronik, T. Stein, S. Refaely-Abramson, and R. Baer. Excitation gaps of finite-sized systems from optimally tuned range-separated hybrid functionals. *J. Chem. Theory Comput.*, vol. 8, no. 5, pp. 1515–1531 (2012).

[30]   L. Goerigk and S. Grimme. Double-hybrid density functionals provide a balanced description of excited 1La and 1Lb states in polycyclic aromatic hydrocarbons. *J. Chem. Theory Comput.*, vol. 7, no. 10, pp. 3272–3277 (2011).

[31] L. Goerigk, A. Hansen, C. Bauer, S. Ehrlich, A. Najibi, and S. Grimme. A look at the density functional theory zoo with the advanced GMTKN55 database for general main group thermochemistry, kinetics and noncovalent interactions. *Phys. Chem. Chem. Phys.*, vol. 19, no. 48, pp. 32184–32215 (2017).

[32] D. Intrieri *et al.*, Indoles from Alkynes and Aryl Azides: Scope and Theoretical Assessment of Ruthenium Porphyrin-Catalyzed Reactions. *Chem. - A Eur. J.*, vol. 25, no. 72, pp. 16591–16605 (2019).

[33] A. Soncini, A. M. Teale, T. Helgaker, F. De Proft, and D. J. Tozer. Maps of current density using density-functional methods. *J. Chem. Phys.*, vol. 129, no. 7 (2008).

Nano Hybrids and Composites
ISSN: 2297-3370, Vol. 33, pp 47-60
© 2021 Trans Tech Publications Ltd, Switzerland

Submitted: 2021-04-27
Revised: 2021-08-22
Accepted: 2021-08-23
Online: 2021-10-11

Liquid-Phase Exfoliation of Graphene in Organic Solvents with Addition of Picric Acid

Syed Sajid Ali Shah[1,3], Habib Nasir[2*] and Shehla Honey[3]

[1]School of Chemical and Materials Engineering, National University of Sciences and Technology, Islamabad, Pakistan

[2]School of Natural Sciences, National University of Sciences and Technology, Islamabad, Pakistan

[3]Centre for Nanosciences, University of Okara, Okara, Punjab, Pakistan

[1]sajidalishah15@gmail.com, [2]habibnasir@sns.nust.edu.pk, [3]shehlahoney@yahoo.com

Keywords: Graphene, sonication, exfoliation, AFM, SEM, and picric acid

Abstract. In this work, graphene was produced by liquid-phase exfoliation of graphite in different organic solvents with the addition of picric acid. The graphene was easily produced by one-step ultra-sonication of graphite powder in the organic solvents. The addition of picric acid has increased the graphene production yield in most of the solvents tested in this work. Picric acid serves as a "molecular wedge" to intercalate into the edge of graphite, which plays a key role during sonication and significantly improves the production yield of graphene. The products were analyzed by microscopic techniques, including atomic force microscopy (AFM) and scanning electron microscope (SEM). The AFM images indicate that the exfoliation efficiency and amount of graphene increased by the addition of picric acid in organic solvents. Moreover, the AFM images also indicate the presence of bilayer graphene. SEM analysis also shows that the addition of picric acid into the organic solvent favors the exfoliation process. The produced graphene was also analyzed by XRD, FTIR, Raman, and UV-visible spectroscopy. The XRD results illustrate that exfoliation was best achieved in N-methyl-2-pyrrolidone (NMP) as a solvent. FTIR and Raman results indicate that the addition of picric acid has slightly defected the produced graphene surface. The amount of graphene concentration was calculated by using Beer-Lambert law, and it was observed that the graphene production yield was increased by using picric acid in most of the solvents. The maximum amount of graphene concentration (0.159 mg/ml) was achieved by adding 30 mg of picric acid in NMP.

1.   Introduction

Graphene has a layered structure in which carbon atoms are linked together through sp^2 hybridization [1, 2]. It has excellent thermal, electrical, and mechanical properties [3]. Therefore, it is used in various applications [4]. It is prepared by different methods including mechanical exfoliation [5, 6], chemical exfoliation [7, 8], electrochemical exfoliation [9], and chemical vapor deposition [10-12]. In all these methods chemical exfoliation method is mostly used to produce a large quantity of graphene. The chemical exfoliation method has two routes, first oxidation-reduction method and the second liquid phase exfoliation method [13-15]. The graphene produced by oxidation-reduction method has defects on its surface, therefore structural properties of graphene are not completely retained in this method. Therefore, liquid-phase exfoliation is carried out to produce graphene for its application in various fields [13, 16, 17] In liquid-phase exfoliation the wander forces between the different layers of graphite are reduced by using sonication in a suitable solvent. The solvent molecules are derived between the graphite layer, which results in exfoliation [6, 18-20].

The interaction between the graphite and solvent should be strong enough to overcome the van der Waals forces between the graphitic layers. This method produces defects-free graphene however, the yield of graphene is very low. The high yield can be obtained by using long time sonication, which is unsuitable for commercial application. Therefore, suitable surfactants are used to increase the production yield of graphene [21-24].

The presence of surfactant increases the particle stability and promotes the exfoliation of graphite by decreasing the surface tension of the solution to match the cohesive energy between the graphene sheets. In this manner, various surfactants have been used to prepare stable graphene suspensions [25]. Many groups have used different surfactants for the liquid-phase exfoliation of graphene such as the linear alkanes exposing a carboxylic acid functional head group (fatty acids) [26]. Recently Polycyclic organic hydrocarbons are the best choice for use as a surfactants [18].

In this work, graphene was exfoliated by liquid-phase exfoliation in different solvents. Moreover, picric acid was used to facilitate the exfoliation process. It was noted that the production of graphene was increased by the addition of picric acid. The picric acid enables the solvent to overcome the van der Waals forces present between the adjacent layers of graphite and considerably increases the production yields of graphene.

2.   Experimental Work

2.1   Materials

Natural graphite powder with an average crystal size of 450 micrometers was obtained from Nacional de Graphite, Brazil. All the solvents used in this work and picric acid were obtained from Sigma- Aldrich and used as received. Moreover, picric acid should be kept and used in wet conditions.

2.2   Exfoliation

In a typical method, 0.5 g of graphite powder was dispersed in 100 ml of solvents. In addition, 0.5 g graphite powder was also dispersed in the same solvents containing picric acid. These dispersions were sonicated for 10 hours. For the removal of solvents, the resultant dispersions were centrifuged for 1 hour at 4000 rpm. After centrifugation, the solvents were removed, and the graphene samples were dried in a vacuum oven at 60°C for 36 hours. The schematic diagram of exfoliation of few-layer graphene is shown in figure 1.

Figure 1: Schematic flow diagram of liquid-phase exfoliation of graphene in organic solvents with addition of picric acid.

2.3   Characterizations

The presence of graphene sheets in the products was confirmed by atomic force microscopy (AFM) JSM-5200 (JEOL) with an operating frequency of 174.504 KHz on SiO_2 substrate. Scanning electron microscope (SEM) micrographs were recorded by using a JSM-490A scanning electron microscope produced by JEOL. The micrographs were recorded in gentle beam mode without metallic coating. X-ray diffraction (XRD) patterns were recorded on X-ray diffraction (XRD) Theta-Theta produced by STOE, Germany. UV-Visible spectra of diluted graphene suspension were recorded by using a UV-visible spectrophotometer (UV-2800 UV-Vis spectrophotometer) the USA in a quartz cell with a one-centimeter optical path over the wavelength range of 200 nm to 800 nm. By measuring the absorbance of graphene, the concentration of exfoliated graphene was calculated by using the Beer-Lambert law. The FTIR spectra of graphene were taken by using FTIR (KBr disk method; Perkin Elmer 2000 spectrometer, USA), at the frequency in the range of 400–4000 cm^{-1}.

3.   Results and Discussion

3.1   XRD Analysis

Figure 2 shows the XRD pattern of graphene sheets produced by using different organic solvents. The peak at 26.5° corresponds to the 2-theta diffraction with an interlayer spacing of 3.46 Å. On examination of graphene sheets produced by using different solvents, the (002) peak gets weaker indicating the fewer numbers of layers and high efficiency of exfoliation of the graphene, or in other words, few-layer graphene is formed. As the peak gets stronger, indicating the increased number of graphene layers [27], in other words, we can say that multilayer graphene is exfoliated. So, figure 2 indicates that NMP as a solvent has shown the best result, while xylene as a solvent has shown the least exfoliation. Few-layers graphene produced by using the surfactant in water has a sharp distinct peak at 26.59°, a peak position which corresponds to d-spacing of 3.35 Å. In comparison to GO which have been reported to have d-spacing ranging from 7.85 Å to 8.07 Å due to the presence of functional groups [28].

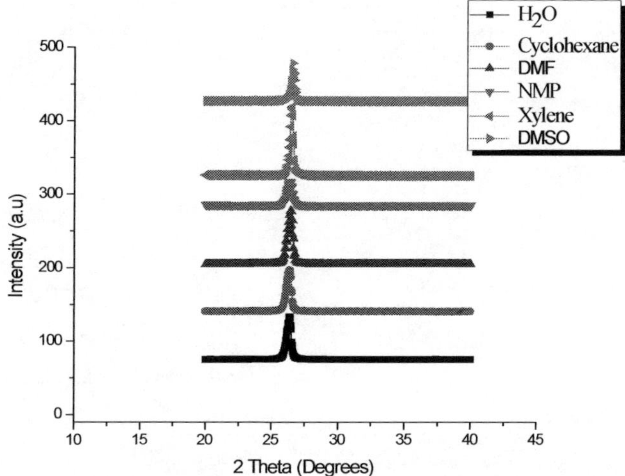

Figure 2: XRD graphs of a few-layers graphene produced liquid-phase exfoliation in different solvents with the addition of picric acid.

3.2 AFM

For calculating the number of layers of graphene from the AFM image, the environmental conditions and substrate are considered. For mica as a substrate, the thickness of a single-layer of the graphene is approximately 0.40 nm while for using SiO_2 as the single-layer graphene thickness is approximately 1 nm thick [18]. The production of graphene by liquid-phase exfoliation of graphite using organic solvents may result from the presence of organic solvent between the layers of graphene. Therefore, the size of organic solvent should be considered in measuring the thickness of graphene sheets. Figure 3(A) shows the AFM image of graphene sheets produced by liquid-phase exfoliation by using picric acid and has the thickness of 2.6 nm, which shows the proof of bilayer graphene with some adsorbed organic solvent. Figure 3(B) shows AFM images of graphene sheets produced without using picric acid and showed the thickness of 4.22 nm, which should be four-layer graphene because a single layer sheet has AFM thickness of ca. 1 nm [29].

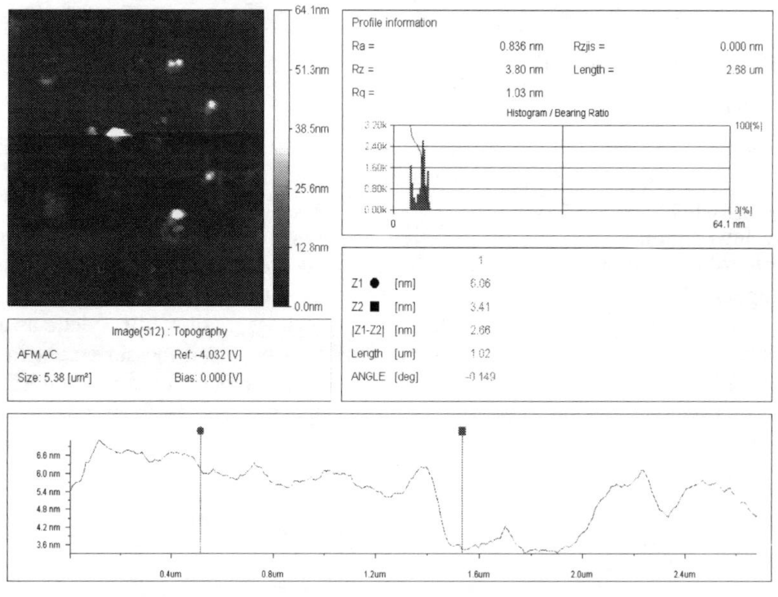

A

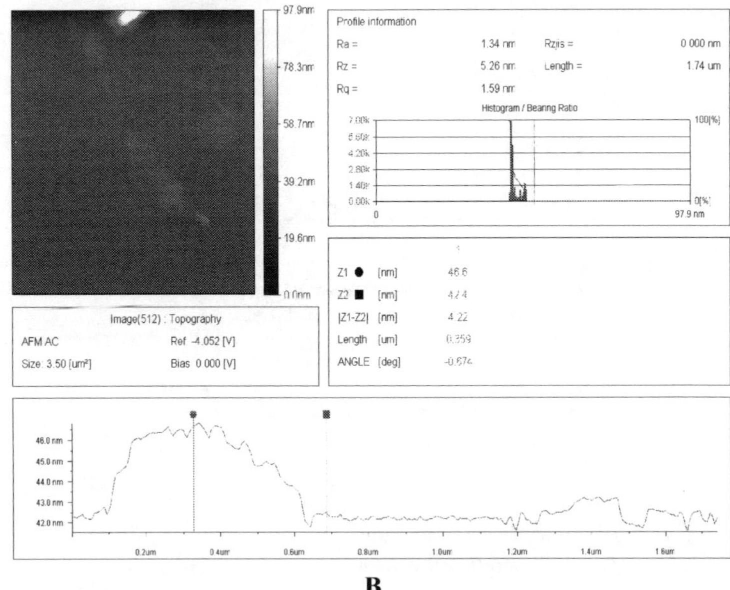

B

Figure 3: (A) AFM analysis of graphene produced with the addition of picric acid and B) AFM analysis of graphene produced without using the picric acid.

3.3   FTIR and Raman Spectroscopy

To study the defects on the graphene surface FTIR and Raman measurements were conducted. Figure 4 shows the FTIR spectra of graphite, graphene produced without using the picric acid and graphene produced by the addition of the picric acid. It is clear from figure 4 that starting material graphite was free from edge defects. Small edge defects were confirmed by FTIR spectra of graphene produced without using picric acid. The use of picric acid has slightly increased the edge defects on the graphene surfaces. The peak at 3450 cm^{-1} ascribed to OH bond stretching vibration of adsorbed water molecules. The peak at 1650 cm^{-1} is due to OH bending vibrations and the peak at 1560 cm^{-1} reflects the vibration of graphene skeletal [30-32].

Figure 5 shows the Raman spectra of graphite, graphene produced by using picric acid and graphene produced without using picric acid. Figure 5 indicates that there are two peaks in the range of 1200 cm^{-1} to 1700 cm^{-1}, characteristically at around 1350 cm^{-1} is the characteristic peak of the D band is due to structural imperfections, thus produced graphene has some defects on its surface. The peak at around 1560 cm^{-1} is the characteristic peak of the G band is ascribed to sp^2 bonded carbon atoms, the peak around 2650 cm^{-1}, the 2D band is due to low phonon double resonance [33, 34]. The 2D band is also denoted by the G' in some work [35], as it is the 2nd prominent band of graphite structure along with the G band [33]. In this work, we denote it as the 2D band to highlight the fact that it is the 2nd order overtone of D band. The main difference between the Raman features of graphene and graphite is the 2D band. For graphene, the 2D band has a symmetric peak while the graphite has no symmetric peak [34]. Figure 5 also indicates that the 2D band becomes broader in the case of graphene produced without using picric acid, shows the increased thickness of graphene [34]. However, the use of surfactant decreases the thickness of graphene as shown in a 2D band. However, the use of picric acid increases the defects on the surface of graphene as indicated by the D band [31]. The 2D band of bilayer and few-layer graphene are not unambiguous in the Raman spectra, therefore bilayer graphene may be produced by using picric acid as determined by the AFM. The ratio of D band to G band (I_D/I_G) is found to be 0.80 and 0.83 for graphene samples produce by using without and with picric acid, respectively. The intensity ratio shows that the use of picric acid has slightly increased the defects on the surface of graphene [34, 36].

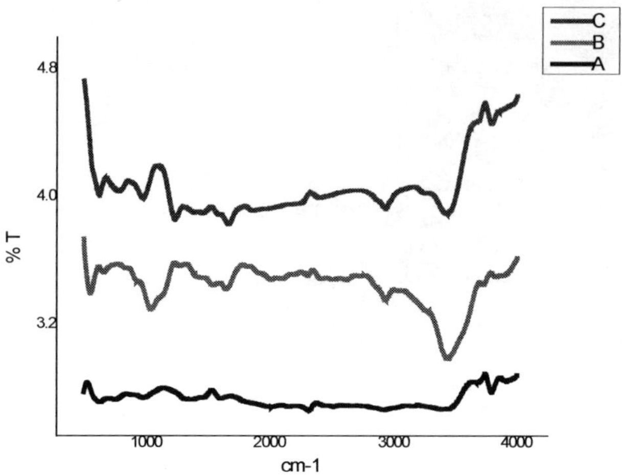

Figure 4: FTIR Spectra of (A) graphite (B) graphene produced with addition of the picric acid (C) graphene produced without using the picric acid.

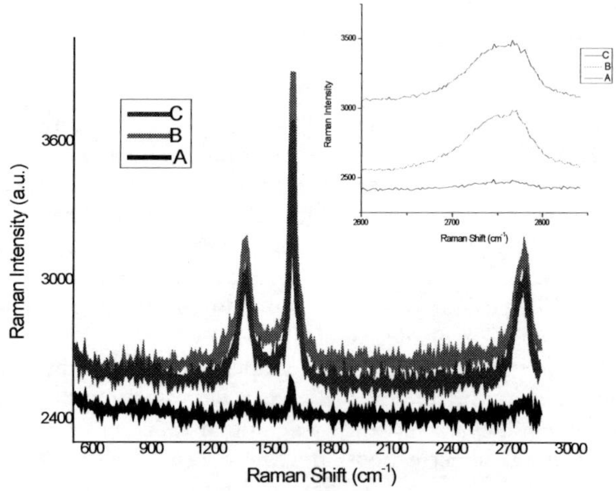

Figure 5: Raman Spectra of (A) graphite (B) graphene produces with addition of picric acid (C) graphene produced without using the picric acid.

3.4   SEM

Figure 6 shows the SEM micrographs of graphite and graphene produced by liquid-phase exfoliation. The morphology of graphene sheets produced by using different solvents was like each other. Figure 6a shows the SEM image of graphite and indicated that the starting material graphite had a layered structure and the layer widths remained much high [37]. Figure 6b illustrates the SEM image of few-layer graphene produced by one-step sonication. The results show that the one-step sonication leads to the exfoliation of graphite. Figure 6c and 6d illustrate the SEM images of few layers' graphene produced by sonication in addition of the use of picric acid as a surfactant and illustrate that use of surfactant had facilitated the exfoliation. Figure 6e and 6f show the SEM images of a few-layers graphene produced by one step sonication by using an extra amount of picric acid as a surfactant and indicate that the extra amount of picric acid remains on the surface of graphite, so, restricts the exfoliation of graphene.

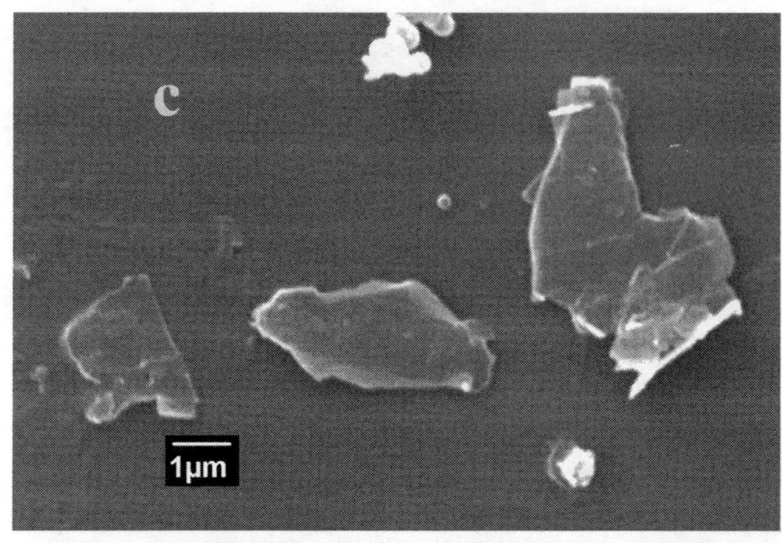

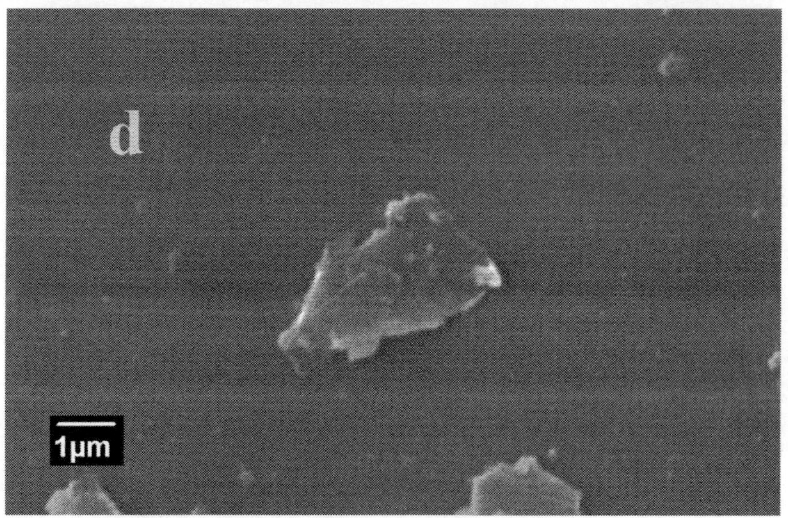

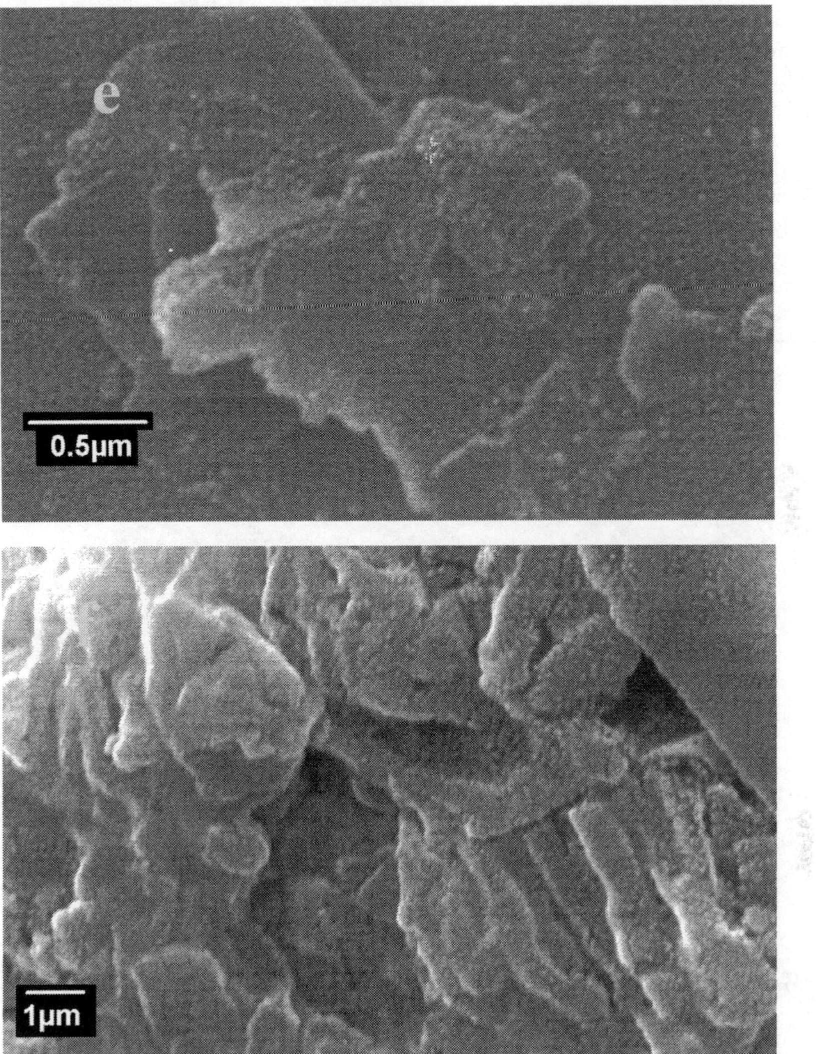

Figure 6: (a) SEM image of graphite (b) and (c) SEM image of few-layer graphene produced by one step sonication, (d) SEM image of few-layer graphene produced by sonication in addition to the use of picric acid, (e) and (f) SEM image of few layers' graphene produced by one step sonication by using an extra amount of picric acid.

3.5   Graphene Production yield

To calculate the concentration of the exfoliated graphene in different solvents, we measured the UV-visible absorbance spectra of the graphene samples produced by different solvents using with and without picric acid, and calculated graphene concentration based on Lambert-Beer law that has been widely used to determine the concentration of graphene dispersed in different solvents [38, 39]. For graphene concentration calculation, the absorbance of graphene was measured at 660 nm, the absorbance coefficient (α), is related to absorbance per unit length, A/l. According to Lambert-Beer law, the relationship between absorbance, absorbance coefficient, and concentration is given by the following equation:

$$A/l = \alpha C$$

Where A is absorbance, l is the length of the cuvette (1 cm), α is absorbance coefficient and C is the concentration of graphene in the dispersion. The value of α was experimentally determined as calculated by Min Yi et al. for different solvents [40-43]. To determine α we prepared five graphene dispersion samples whose concentration was known. We obtained the precise relationship between the A/l and graphene concentration as shown in figure 7. The absorbance coefficient was obtained by a straight line at 660 nm wavelength. For NMP it was taken 2465 ml/(mg.m) and the reported value is 2460 ml/(mg.m) [31]. For water it was 3590 ml/(mg.m), which is very close to reported 3600 ml/(mg.m) [40] and for DMF calculated α is 3460 and reported is 3463 [44]. It is clear from figure 8 that the addition of picric acid up to 30mg/100ml has increased the graphene production yields, however when the surfactant concentration was above the 30mg /100ml, then it has an adverse effect on the graphene production yields. The highest graphene yield production (0.159 mg/ml) was achieved by using NMP as a solvent with the addition of 30 mg picric acid as surfactant. When surfactant concentration is exceeded above 30 mg then it caused adverse effects on the exfoliation process. Figure 9 shows the concentration of dispersed graphene in solvents using with or without the addition of picric acid as a surfactant and revealed that the addition of picric acid has doubled the graphene production yields in most of the solvents. The graphene production yield was nearly doubled due to the addition of picric acid.

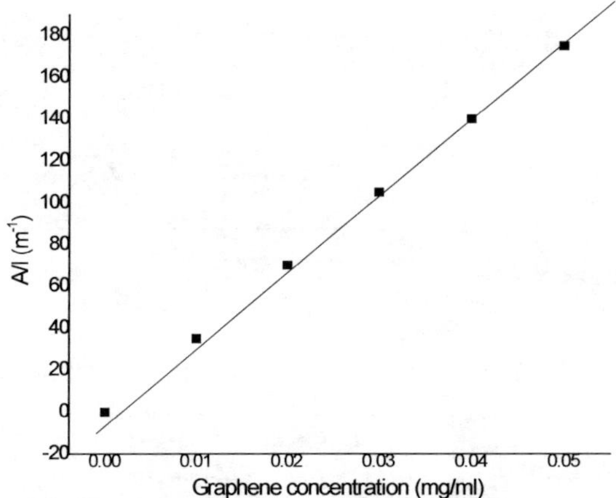

Figure 7: Graph shows the linear relationship between absorbance per unit length (A/C) and graphene concentration

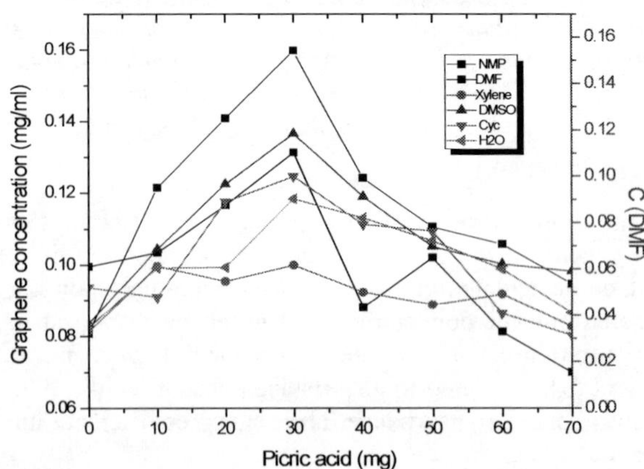

Figure 8: Effect of concentration of picric acid on the production of graphene in different solvents.

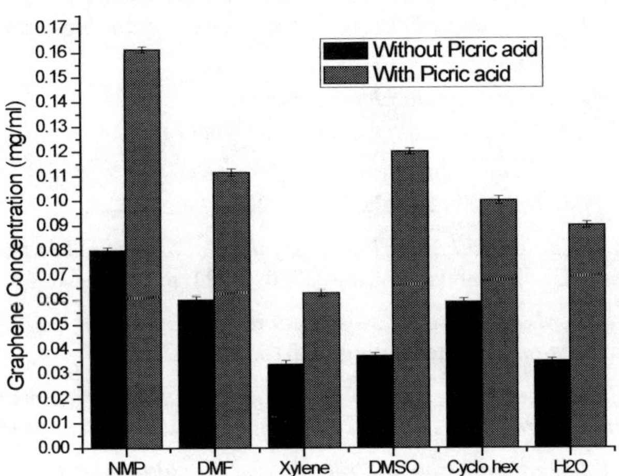

Figure 9: Graph shows the graphene concentration dispersed in different solvents with and without the addition of the picric acid.

4.   Conclusion

The careful choice of solvent that has surface energy matching with that of graphene can capably exfoliate graphite to directly get graphene. We produce graphene by one step liquid-phase exfoliation of the graphite in organic solvents using picric acid. Picric acid plays a vital role in exfoliation and assist as a "molecular wedge" to intercalate on the edge of the graphite. The UV-visible spectroscopic results supported that; the use of picric acid has doubled the graphene production in most of the solvents. The exfoliation of graphite powder under ultrasonic energy enables the organic solvent to efficiently penetrate through the interlayer spacing of graphite with diffusivity and much more energy than the interlayer energies of graphite. The AFM images show that few-layer graphene is produced with notable content of bilayer graphene sheets. SEM and XRD also confirmed the exfoliation of graphene. Graphene produced in this work retains the original pristine structure of graphite without the crystal defects.

Acknowledgment

This work was supported by the Research Fund of the National University of Sciences and Technology, Islamabad, Pakistan.

References

[1]   Chia, J.S.Y., et al., *Facile synthesis of few-layer graphene by mild solvent thermal exfoliation of highly oriented pyrolytic graphite.* Chemical Engineering Journal, 2013. 231: p. 1-11.

[2]   Qiu, X., et al., *Liquid-phase exfoliation of graphite into graphene nanosheets in a hydrocavitating 'lab-on-a-chip'.* RSC Advances, 2019. 9(6): p. 3232-3238.

[3]   Castarlenas, S., et al., *Few-layer graphene by assisted-exfoliation of graphite with layered silicate.* Carbon, 2014. 73: p. 99-105.

[4]   Zhang, L., et al., *Rationally Designed Surfactants for Few-Layered Graphene Exfoliation: Ionic Groups Attached to Electron-Deficient pi-Conjugated Unit through Alkyl Spacers.* Acs Nano, 2014. 8(7): p. 6663-6670.

[5]   Ritter, K.A. and J.W. Lyding, *Characterization of nanometer-sized, mechanically exfoliated graphene on the H-passivated Si(100) surface using scanning tunneling microscopy.* Nanotechnology, 2008. 19(1).

[6]   Niu, Y., et al., *Mechanical and liquid phase exfoliation of cylindrite: a natural van der Waals superlattice with intrinsic magnetic interactions.* 2D Materials, 2019. 6(3): p. 035023.

[7]   Al-Hazmi, F.S., et al., *One pot synthesis of graphene based on microwave assisted solvothermal technique.* Synthetic Metals, 2015. 200: p. 54-57.

[8]   Paolucci, V., et al., *Sustainable Liquid-Phase Exfoliation of Layered Materials with Nontoxic Polarclean Solvent.* ACS Sustain Chem Eng, 2020. 8(51): p. 18830-18840.

[9]   Huang, X.H., et al., *Low defect concentration few-layer graphene using a two-step electrochemical exfoliation.* Nanotechnology, 2015. 26(10).

[10]   Struzzi, C., et al., *High-quality graphene on single crystal Ir(111) films on Si(111) wafers: Synthesis and multi-spectroscopic characterization.* Carbon, 2015. 81: p. 167-173.

[11]   Cui, W., et al., *A Strong Integrated Strength and Toughness Artificial Nacre Based on Dopamine Cross-Linked Graphene Oxide.* Acs Nano, 2014. 8(9): p. 9511-9517.

[12]   Zhao, Y., et al., *Direct growth of graphene on gallium nitride by using chemical vapor deposition without extra catalyst.* Chinese Physics B, 2014. 23(9).

[13]   Chua, C.K., et al., *Chemical Preparation of Graphene Materials Results in Extensive Unintentional Doping with Heteroatoms and Metals.* Chemistry-a European Journal, 2014. 20(48): p. 15760-15767.

[14]   Mishra, A.K. and S. Ramaprabhu, *Functionalized graphene sheets for arsenic removal and desalination of sea water.* Desalination, 2011. 282: p. 39.

[15]   Mukhopadhyay, T.K. and A. Datta, *Disentangling the liquid phase exfoliation of two-dimensional materials: an "in silico" perspective.* Physical Chemistry Chemical Physics, 2020. 22(39): p. 22157-22179.

[16]   Park, J.S., et al., *Liquid-phase exfoliation of expanded graphites into graphene nanoplatelets using amphiphilic organic molecules.* Journal of Colloid and Interface Science, 2014. 417: p. 379-384.

[17]   Masoumi, Z., M. Tayebi, and B.K. Lee, *Ultrasonication-assisted liquid-phase exfoliation enhances photoelectrochemical performance in alpha-Fe2O3/MoS2 photoanode.* Ultrason Sonochem, 2021. 72: p. 105403.

[18]   Ciesielski, A. and P. Samori, *Graphene via sonication assisted liquid-phase exfoliation.* Chemical Society Reviews, 2014. 43(1): p. 381-398.

[19]   Alaferdov, A.V., et al., *Size-controlled synthesis of graphite nanoflakes and multi-layer graphene by liquid phase exfoliation of natural graphite.* Carbon, 2014. 69: p. 525-535.

[20]   Hu, C.X., et al., *Dispersant-assisted liquid-phase exfoliation of 2D materials beyond graphene.* Nanoscale, 2021. 13(2): p. 460-484.

[21]   Georgakilas, V., et al., *Hydrophilic Nanotube Supported Graphene-Water Dispersible Carbon Superstructure with Excellent Conductivity.* Advanced Functional Materials, 2015. 25(10): p. 1481-1487.

[22]   Guardia, L., et al., *High-throughput production of pristine graphene in an aqueous dispersion assisted by non-ionic surfactants.* Carbon, 2011. 49(5): p. 1653-1662.

[23]   Liang, B., et al., *Organic salt-assisted liquid-phase shear exfoliation of expanded graphite into graphene nanosheets.* Journal of Materiomics, 2021.

[24] Gu, X., et al., *Method of ultrasound-assisted liquid-phase exfoliation to prepare graphene.* Ultrason Sonochem, 2019. 58: p. 104630.

[25] Sham, A.Y.W. and S.M. Notley, *Layer-by-Layer Assembly of Thin Films Containing Exfoliated Pristine Graphene Nanosheets and Polyethyleneimine.* Langmuir, 2014. 30(9): p. 2410-2418.

[26] Haar, S., et al., *A Supramolecular Strategy to Leverage the Liquid-Phase Exfoliation of Graphene in the Presence of Surfactants: Unraveling the Role of the Length of Fatty Acids.* Small, 2015. 11(14): p. 1691-1702.

[27] Kumar, R., et al., *Pressure-dependent synthesis of high-quality few-layer graphene by plasma-enhanced arc discharge and their thermal stability.* Journal of Nanoparticle Research, 2013. 15(9).

[28] Chia, J.S.Y., et al., *Facile synthesis of few-layer graphene by mild solvent thermal exfoliation of highly oriented pyrolytic graphite.* Chemical Engineering Journal, 2013. 231(0): p. 1-11.

[29] Xu, L.X., et al., *Production of High-Concentration Graphene Dispersions in Low-Boiling-Point Organic Solvents by Liquid-Phase Noncovalent Exfoliation of Graphite with a Hyperbranched Polyethylene and Formation of Graphene/Ethylene Copolymer Composites.* Journal of Physical Chemistry C, 2013. 117(20): p. 10730-10742.

[30] Feng, Y., et al., *Degradation of 14C-labeled few layer graphene via Fenton reaction: Reaction rates, characterization of reaction products, and potential ecological effects.* Water Research, 2015. 84: p. 49-57.

[31] Xu, J.S., et al., *Liquid-phase exfoliation of graphene in organic solvents with addition of naphthalene.* Journal of Colloid and Interface Science, 2014. 418: p. 37-42.

[32] Tang, Z., J. Zhuang, and X. Wang, *Exfoliation of Graphene from Graphite and Their Self-Assembly at the Oil–Water Interface.* Langmuir, 2010. 26(11): p. 9045-9049.

[33] Ni, Z., et al., *Raman spectroscopy and imaging of graphene.* Nano Research, 2010. 1(4): p. 273-291.

[34] Ferrari, A.C., et al., *Raman spectrum of graphene and graphene layers.* Phys Rev Lett, 2006. 97(18): p. 187401.

[35] Cançado, L.G., et al., *Geometrical approach for the study of G'band in the Raman spectrum of monolayer graphene, bilayer graphene, and bulk graphite.* Physical Review B, 2008. 77(24).

[36] Mattevi, C., et al., *Evolution of Electrical, Chemical, and Structural Properties of Transparent and Conducting Chemically Derived Graphene Thin Films.* Advanced Functional Materials, 2009. 19(16): p. 2577-2583.

[37] Güler, Ö. and S.H. Güler, *Production of graphene–boron nitride hybrid nanosheets by liquid-phase exfoliation.* Optik - International Journal for Light and Electron Optics, 2016. 127(11): p. 4630-4634.

[38] Liu, W.W., et al., *Exfoliation and dispersion of graphene in ethanol-water mixtures.* Frontiers of Materials Science, 2012. 6(2): p. 176-182.

[39] Arifutzzaman, A., et al., *Experimental investigation of concentration yields of liquid phase exfoliated graphene in organic solvent media.* IOP Conference Series: Materials Science and Engineering, 2019. 488: p. 012001.

[40] Yi, M., et al., *Achieving concentrated graphene dispersions in water/acetone mixtures by the strategy of tailoring Hansen solubility parameters.* Journal of Physics D: Applied Physics, 2013. 46(2): p. 025301.

[41] Konios, D., et al., *Dispersion behaviour of graphene oxide and reduced graphene oxide*. J Colloid Interface Sci, 2014. 430: p. 108-12.

[42] Wang, S., M. Yi, and Z. Shen, *The effect of surfactants and their concentration on the liquid exfoliation of graphene*. RSC Adv., 2016. 6(61): p. 56705-56710.

[43] Singh, R. and C.C. Tripathi, *Enhancing Liquid-Phase Exfoliation of Graphene with Addition of Anthracene in Organic Solvents*. Arabian Journal for Science and Engineering, 2017. 42(6): p. 2417-2424.

[44] Shang, J., F. Xue, and E. Ding, *The facile fabrication of few-layer graphene and graphite nanosheets by high pressure homogenization*. Chem. Commun., 2015. 51(87): p. 15811-15814.

Nano Hybrids and Composites
ISSN: 2297-3370, Vol. 33, pp 61-72
© *2021 Trans Tech Publications Ltd, Switzerland*

Submitted: 2021-06-01
Revised: 2021-09-03
Accepted: 2021-09-10
Online: 2021-10-11

New Results on Diffusion in Graphene Nanostructures for Sensoristics

Paolo Di Sia[1,a]

[1]University of Padova, Department of Physics and Astronomy & Department of Chemical Science,
Via Marzolo 8, 35131 Padova, Italy

[a]paolo.disia@gmail.com

Keywords: Graphene, Nanostructures, Sensoristics, Advanced materials, Diffusion, Modelling, DS model.

Abstract. Graphene has particularly interesting chemical and physical properties, including high chemical and mechanical resistance, excellent thermal and electric transport, high transparency. Micro/nanoelectronics represents one of the key enabling technologies (KETs) of the future; it is the basis of innovation and competitiveness of almost all scientific and applicative sectors. Activities involving it are aimed at the development of new materials, processes, devices and technologies in a wide range of sectors. In the field of sensoristics, it is possible to create devices for applications in most sectors of global interest. The paper provides an interesting overview of the possible applications of graphene in relation to its properties, so as interesting details by last efforts in theoretical nanophysics. A recent analytical Drude-Lorentz-type model is able to well describe the conductors in nanostructured form, going to mimic the infrared properties of oxides and semiconductors in the nano form and offering interesting previsions.

Introduction

Throughout history, the technological development has measured the progress and marked the succession of economic and social changes in the civil, scientific and military community. Each era takes the name by an indispensable material, whose discovery or exploitation indisputably marks a change, sometimes radical, in the structure and socio-economic architecture.

Graphene marked the beginning of a new era with the name of one of the most common chemical elements, the carbon, a fundamental element in the technological evolution of human being (Fig. 1).

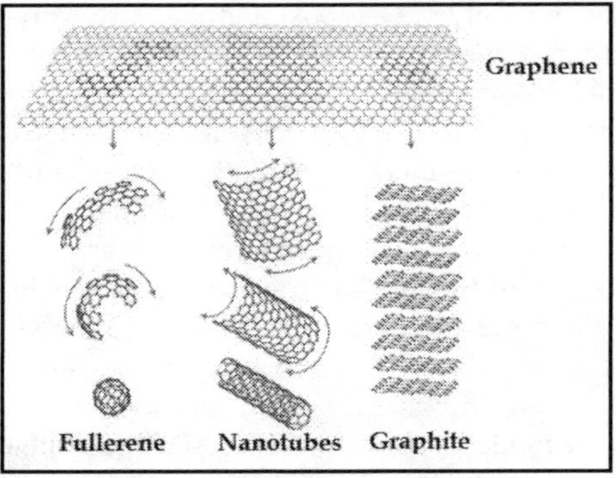

Figure 1. Graphene.

In addition to the generally known macroscopic forms, such as graphite or diamond, carbon is able to form very different nanostructures, with particular physical properties that they can be described as points, two-dimensional or three-dimensional structures. The formation of so different structures with so different physical-chemical properties is to be attributed to the different types of chemical bond with which carbon atoms are able to bind to each others [1]. The atomic honeycomb structure of the crystal lattice gives to graphene unique electronic properties.

The conical shape of the energy bands of graphene makes the charge carriers in this crystal as particles of practically zero effective mass, free to move in the lattice without undergoing significant interactions with the atoms. These characteristics make graphene theoretically one of the materials with the highest electronic mobility existing in nature.

At experimental level, graphene differs from these theoretical limits, since the material analyzed in laboratory has often reticular defects, and typically the processes necessary for its synthesis leave different chemical contaminations on its surface, making the produced material significantly different from the viewpoint of chemical or physical properties. Despite this, there are to date numerous experimental results that have allowed to verify many of the theoretical expectations.

Since the first experiments in 2004, graphene has been isolated in the form of sheets supported on silicon substrates covered with silicon oxide. It was immediately apparent that the presence of the silicon oxide substrate, often inevitably covered by a thin film of contaminants (water, hydrocarbons from the atmosphere, etc.) that interact with the carbon atoms of the graphene lattice, can significantly modify its properties.

In this experimental configuration, graphene has however been characterized for its electronic transport properties, demonstrating ambipolar conduction, i.e. by electrons and holes [2], with electronic mobility up to 10^5 cm^2/Vs [3] in conditions of high vacuum on graphene membranes suspended between two electrodes, i.e. isolated by the substrate; as comparison, the value of silicon used in common electronic devices hardly exceeds 10^3 cm^2/Vs.

Compared to silicon, the mobility of carriers in graphene is practically independent of their concentration and doping level, essential characteristics in the creation of systems capable of high switching speed, of supporting high currents or needing to dissipate high powers [4].

In materials, good electrical conduction is commonly associated, as in the case of metals, with good thermal conduction; in fact, charge carriers are also responsible of mediating and transporting the collective vibrations of the crystal lattice, linked to temperature variations and therefore to thermal excitation, from an end of the material to the other. Numerous experimental data confirmed the excellent thermal conduction of graphene. Its thermal conductivity at room temperature measures between $(4.84 \pm 0.44) \cdot 10^3$ W/mK and $(5.30 \pm 0.48) \cdot 10^3$ W/mK, greater values than those measured for carbon nanotubes ($3.5 \cdot 10^3$ W/mK) or for copper ($0.401 \cdot 10^3$ W/mK) [5].

With regard to mechanical properties, a graphene sheet behaves similarly to a sheet of paper: when we try to pull it along the plane, it offers considerable resistance, but a small force is enough to bend it on itself. Graphene is in fact not only very robust, but it is also extremely light, with a density of only 0.77 mg/m^2 (1 m^2 of normal paper weighs about 1000 times more) and the structure of the lattice facilitates flexion. Graphene is therefore a very resistant and very interesting material for optoelectronic applications, in particular as transparent and conductive electrode.

Graphene is a flexible conductor. For this feature, both in the scientific research and in the private sector, efforts are being made for overcoming the current technical limitations, to produce samples that are gradually more pure and larger. Once these limitations are overcome, graphene can replace ITO and be used in the production of flexible and transparent screens [6-8].

Graphene symbolizes at the same time a challenge and a promise for the future, for material industry and chemical industry, but also for flexible electronics (sensoristics that can be integrated directly into clothing).

A recent Drude-Lorentz-type model is able to well describe the conductors in nanostructured form, going to mimic the infrared properties of oxides and semiconductors at the nano level. Its analytical formulation keeps it powerful and elegant, able also to offer interesting previsions.

Overview on Micro-Nanoelectronics and Sensoristics of Next Future

Micro-nanoelectronics represents one of the key enabling technologies (KETs) of the future and is essential for products and services, being at the basis of innovation and competitiveness. Related activities are aimed at the development of new materials, processes, devices and technologies in the following sectors:

- *Nanoscale nanoelectronic devices* with new generation logic and/or memory functionality, based on emerging concepts, quantum information manipulation and advanced nanofabrication technologies;
- *Unconventional electronic systems* towards the development of multifunctional platforms and innovative computational architectures, such as neuromorphic networks and quantum computation;
- *Enabling technologies* for new generation power and high frequency devices based on advanced materials (SiC, GaN, GaAs, graphene, etc.);
- *Electronic devices and circuits* on flexible substrates based on both organic transistors (OTFTs), made using fully printed technology, and on polycrystalline silicon TFT;
- *Devices based on materials with reduced dimensions* (2-D, 1-D, 0-D) for low-power electronics and flexible electronics, and development of enabling technologies for nanoscale electronic devices;
- *Sensors for environmental surveillance*, suitably functionalized process monitoring (fiber Bragg gratings), sensors distributed by means of fiber optic elements, multisensory systems (electronic noses);
- *Plasmonic optical transduction sensors and biosensors* for quantum sensing using new materials with magneto-optical properties or with specific functions, up to the detection of a single molecule;
- *Development of specific transducers*, based on advanced sensitive materials (quantum dots, nanowires, etc.) and with innovative transduction techniques for high performance sensors;
- *Autonomous multisensory systems* of chemical and physical parameters, also on unmanned platforms (drones);
- *Multifunctional and multisensory systems* for assisted living, safety, prevention and health protection;
- *Flexible sensory systems* with particular reference to wearable devices, including energy generating devices, circuits and flexible devices based on innovative multifunctional materials (oxides, dichalcogenides);
- *Multifunctional systems for medical applications*, such as biosensors, nanomechanical sensors for biological systems, micro-dispensing of drugs, medical diagnosis systems (breath analysis, PET analysis, myocardial heart attack), MOEMS on optical fiber for medical diagnostics;
- *MEMS devices and sub-systems* for microwave and millimeter waves telecommunications such as switches, resonators, micro-nanoantennas;
- *Advanced MEMS/MOEMS technologies* for resonant deformation sensors, energy micro-harvesting, pressure, inertial, acoustic and bolometric sensors, including the integration of innovative materials for piezoelectric, thermal, thermoelectric and chemical transduction;
- *Micro-nanoactuators and resonators* based on unconventional materials (phase change materials);
- *Devices for energy conversion* operating through photo-thermoionic, thermoionic and thermoelectric processes and based on the development of unconventional materials and nanostructures (nanowires);
- *Integrated gravimetric-electrochemical sensors* for the dosage of pathogenic microorganisms, biomarkers, organic pollutants;
- *Computational models of artificial intelligence* for vision and pattern recognition in sectors of security and intelligent surveillance;
- *Memristive systems* with memory functionality, non-volatile logic and for neuromorphic computation systems, such as non-volatile PCM nanoscale memories, trimmable resistors, resistive switching devices (RRAM) based on oxides for memories and as functional elements in neural and neuromorphic architectures, MTJs (Magnetic Tunnel Junctions);
- *Quantum computing*: simulation/modelling and characterization of qubits in compatible CMOS nanodevices;
- *Development of conventional* (gravure, inkjet, screen printing) *and innovative* (piezoelectric and pyroelectric) *printing techniques* for the realization of fully printed devices;
- *Development of sensors and optoelectronic platforms* for harsh environments (high temperature, intense neutron radiation and/or ionizing radiation, chemically aggressive environments, etc.);
- *Biosensors* for monitoring environmental pollutants in the liquid and gas phase, smart bioelectronic systems for depuration and purification of waste water.

Sensing Property of Graphene

In the wide sensing sector, especially in the field of environmental monitoring [9], a particularly important class is that of chemical sensors, in which the graphene layer plays the role of active medium. Already the first experimental studies [10] indicated the potential of graphene as very sensitive material to gases, given that devices were doped following contact with water or exposure to ammonia (NH_3) or ethanol vapors (Fig. 2).

Subsequent theoretical and experimental works confirmed and strengthened this idea; since 2007, chemical sensors capable of detecting a single gas molecule have been created [11]. The ultimate goal of each detection method is in fact to reach a level of sensitivity that can discriminate the "quantum" of the measured quantity. In the case of chemical sensors, the fundamental entity is represented by the atom or molecule of the analyzed species; if this resolution constituted an insurmountable limit even for the most sensitive detectors, the combination of some properties of graphene allows to obtain performance not found in other types of solid state sensors.

The main limitation to the resolution for these types of sensors is generally represented by the intrinsic defects of the material of which they are made, determining a high noise and consequently a low signal-to-noise ratio (SNR) [12].

Other sources, due to the contacts with which the sensor is made, must also be added to the intrinsic noise. The SNR parameter is therefore taken as a factor of merit because the operating principle of these devices is based on the transduction of the adsorption or desorption of gas in the variation of an electrical quantity [13]. In the event that the information is converted into voltage or current, the variation in resistance can be directly obtained and the sensors are also called "chemoresistors".

Figure 2. Flexible and transparent graphene-based electrode.

The graphene flake, acting as a resistor, is subjected to the voltage applied to its ends and the current flows through it via metal contacts. In general, the sensing properties of graphene and the particular application to the field of chemical sensors are mainly explained for reasons related to the particular structure and characteristics of the material.

Various properties of graphene may be interesting to satisfy the previous requests:

- *high chemical stability* of the 2D lattice (unlike CNTs, local defects have little influence on transport and mechanical properties), but also possibility of functionalization (to improve the specificity of the response);

- *doping effect* at the basis of the semiconductor sensors: it makes possible to strongly vary the number of carriers, depending on the chemical species adsorbed on the surface.

Favorable electronic and mechanical properties for implementing transduction mechanisms (e.g. electrical contacts, transistors, etc.) are:

- *high mobility* (speed of response);

- *large available surface* (exposure is maximum, such as adsorption);

- *high conductivity* and *rare defects* (help in reducing the noise and improving SNR);

- possibility of *ohmic contacts* (use with various transducers).

There is the possibility of application also in physical sensors with greater simplicity and lower production costs of graphene (or of its subsequently reduced oxidized forms) compared to CNTs.

- *Semiconductor sensors as gas sensors*: the operating principle is based on the charge transfer by the analyte molecules and requires that a chemisorption takes place on the surface of the semiconductor. The mechanism is of redox type: there may be reducing agents (electron donors) or oxidants (electron acceptors). The resulting doping effect affects the conductivity of the semiconductor. This can in general be applied to the particular case of graphene, although the latter is a border case of semiconductor with zero gap.

Graphene plans obtained with the GO (Graphene Oxide) reduction method and deposited as sensitive part of conductometric devices, respond to gases such as NO_2 and NH_3 (p- and n-type doping respectively).

It can be observed (common to this class of sensors) that the response improves in high temperature regimes. The response and recovery time becomes usable even if at the expense of the response intensity.

The sensitivity to analytes is different according to the chemical species; the variation of carriers is linear with the concentration of the dopant species. Similar devices can reach the resolution threshold of 1 ppb, given the low noise distinguishing them. The observation of quantized variations in resistivity seems to indicate the limit of single adsorption/desorption events.

- *Sensors with (r)GO* (Reduced Graphene Oxide): sensors based on the GO would be too insulating for the conductometric use. Thus, the presence of functional groups and active defects can be exploited. These sensors tend to have two response regimens, one of rapid adsorption and one (slower) of interaction with chemically more active sites such as vacations and functional groups. They show low noise levels also compared to CNT-based sensors.

- *Gated-FET solutions*: they are sensors in which electronic applications of graphene are exploited. This acts as the active channel of a FET, whose gate is controlled by an electrolyte solution, playing the role of the dielectric. From a chemical viewpoint, the previous doping mechanism is exploited, but the transducer changes.

The gate potential affects the ions present in the electrolyte; this has consequences on the charge transfer: it modulates the balance of H_3O^+/OH^- ions at the interface with graphene, making the device sensitive to the pH of the electrolyte solution. FETs based on non-functionalized graphene are therefore sensitive to ionic equilibria and can be good pH sensors. The relationship between pH and measured conductivity is linear.

The device can also be used to determine minimum quantities in solution of a protein; sensitivity up to tenths of nM are obtained. The changes in conductivity are linear for low concentrations.

- *Studies on electrodes*: there are many studies investigating the behaviour of electrodes in functionalized or non-functionalized graphene, using techniques such as cyclic voltammetry and amperometric response.

These studies confirm the importance of defects or functional groups, in order to obtain more specific responses to a certain analyte and improve its adsorption with a more intense interaction. This is particularly valid for molecules of biological interest, for which selective interaction is usually required [14,15].

Possible Applications of Graphene and its Derivatives

There are many possible applications for graphene and its derivatives:

- *Electrochemical and biological sensors* to detect the presence of gas or changes in biological environments.

- *Mass sensors*.

- *Composite materials with polymers*, light and with high mechanical properties for aerospace uses.

- *Optoelectronic devices*.

- *Absorbent and catalytic materials* for the capture and photodegradation of both organic and inorganic pollutants, both in the liquid and gas phase.

- *Membranes* to increase the TEM resolution.

- *Biomedical applications*: nanomedicine is one of the most interesting applications and one of the most promising developments in nanotechnology. It includes a wide range of research activities, ranging from the development of biosensors, nanomaterials with biomedical applications, to the construction of nanovectors for therapeutic and diagnostic purposes. The nanosize has significant advantages in the biomedical and pharmacological field, where the reduction of volumes to benefit of exchange surfaces is able to significantly improve the interactions between nanomaterials and living cells.

We can use different types of nanoparticles, from polymeric to metallic ones, liposomes, dendrimers, microcapsules, etc; graphene related materials (GRMs) are becoming very important in nanomedicine and biology.

An interesting approach for covalent functionalization of graphene uses GO, providing a new class of polychromatic dispersible solution platform for chemical use. Functional groups make the GO hydrophilic, allowing it to be dispersed in water. Thanks to functional groups, GO can interact with a wide range of organic and inorganic species in ionic, covalent and non-covalent way, making functional hybrids synthesizable.

GO is fluorescent on a wide range of wavelengths. This tunable fluorescence is currently used in biological/medical sensing applications and drug delivery. Graphene and CNTs show a similar behaviour, with graphene providing supplementary functionality compared to carbon nanotubes, such as a greater load of biomolecules, due to its two-dimensional shape.

Current research includes chemical functionalization controlled by graphene with functional units to reach good processability in media and regulation of various physical-chemical properties. A purpose of the controlled surface oxidation is the realization of anchor points for extra surface groups aimed at the union of biomolecules (peptides, DNA, growth factors) by means of carboxylic groups [16-18].

- *Applications related to biosensors and bioelectronics*: GFETs (graphene-based field-effect transistors) are interesting for their potentiality in bioelectronics, particularly as an interface between living cells and nervous tissue. The association of superb electronic properties, bio-chemical stability and compatibility, and easy integration with flexible technology brings graphene to be a competitor for the near future generation of neuroprostheses.

- *Targeted drug delivery*: there is high experimental research for the production and characterization of nanoparticles that can act as efficient vectors for drug delivery, understood as the development of alternative targeted-distribution systems for drugs in the body. This aims to limit the biological effect of therapy to a specific type of cells, improving the efficacy and at the same time reducing the systemic toxicity.

Nanovectors are one of the best alternatives for the administration of medicines to chronically ill patients, who need continuous treatments, usually at high dosages, often involving significant collateral effects.

At the same time, the great development of the techniques of realization and surface modification of materials at the nanometric level has opened new possibilities previously only imaginable, such as for example nanostructured solids to which biomolecules can specifically bind or inert inorganic matrices for encapsulation of biomolecules.

This development is part of the search of alternative methods for the treatment of highly critical diseases, in order to significantly improve the quality of life of patients and their hope of survival.

The study of biointerfaces, surfaces of biomaterials coming into contact with tissues or biological fluids, is essential. Interactions with the vascular system such as blood or blood vessels are critical since they can trigger a variety of events threating the patient's life. In many cases, in fact, the adsorption of proteins on the interface initiates cascade reactions leading to thrombosis or to activation of the immune system and complement, defense mechanisms that respond when infectious agents are recognized in the body.

These reactions are the main responsible for the rejection, therefore the study of the biocompatibility of materials to the biointerfaces is an integral part in the research and design of drug delivery. To minimize the risks related to the use of biomaterials, interfaces are modified so that the system can amalgamate with body fluids; an alternative approach is to make it invisible to the body's defense mechanism. These results direct towards the administration of graphene-based drugs.

- *Imaging and diagnosis*: the QDs luminescences (quantum dots, fluorescent nanocomposites) are widely used for biological labeling (bio-labeling) and bioimaging. But their toxicity and the potential environmental danger have limited their diffusion in in vivo applications. Biocompatible carbon-based fluorescent nanomaterials can result a viable alternative. Fluorescent species in IR and NIR spectroscopy are helpful for bio-medical applications, showing cells and tissues small self-fluorescence in this area.

Graphene-based luminescent materials can be produced by covering IR, visible and blue. The development of graphene-based materials in this sector needs to be compared with the search of possible toxic effects.

- *Dental diseases*: dental diseases, caused by the overgrowth of some bacteria in the mouth, are among the most common health problems in the world. Graphene oxide has been found to be effective in eliminating these bacteria, considering that some of them have developed antibiotic resistance. Studies are showing that graphene oxide can inhibit some bacterial strains with a minimal damage to mammalian cells. By destroying walls and membranes of bacterial cells, graphene oxide slows the growth of pathogens, thus having potential uses in dental care.

- *Joint prostheses*: some applications ask for hydrophobic materials with a non-adhesive surface to the cell, as instruments in contact with human blood or joint prostheses in the attrition zone, in other cases an adhesive surface to the cell to ensure complete integration into the tissue of the implanted material. Graphene is an interesting way as a biocompatible coating, so as other carbon layers like nano-diamond coatings or DLC (Diamond Like Carbon), innovative carbon-based coating.

Graphene is usable also as a strengthening for polymer and ceramic prostheses. Tiny percentages of graphene or GO valorise the elongation at break of the polymer, guiding to harder materials. Graphene-based polymeric compounds exhibit as well good tribological properties.

- *Neural prostheses*: emerging studies on neural prostheses, in the brain, eyes or spinal column, promise the development of technologies capable of fighting degenerative diseases, repairing damaged tissues, strengthening senses.

Silicon, material used to create experimental implants, does not well integrate with the soft tissues of our organism: it is rigid, sharp and risks damaging the substrate that surrounds it. We therefore need a flexible material and more compatible with human tissues, which can be just graphene, thanks to its particular intrinsic properties: thin, very flexible, extremely resistant, very sensitive to electrical variations, therefore ideal for this type of systems.

Physical, Chemical and Biological Sensoristics of Graphene

This sector has devices that, following an external stimulus, change their state and through the variation of measurable properties, allow the interpretation of a signal. They can be distinguished *a priori* according to the field of application: physical, chemical and biological sensors.

Chemical and biological sensors are used for concentration measurements of a chemical species. Generally the analyte is in the gas phase or in solution. The latter case is obligatory with regard to biological sensors, considering that the recognition of the analyte takes place by means of biological substances.

They consist of an active part of the device, that must react to changes in the concentration of the analyte, and a passive part, acting as a transducer of the response in an easily acquired signal (in many cases electrical signals are acquired, therefore there may be successive stages of amplification, conversion, etc.) (Fig. 3).

Depending on the transduction mechanism, devices can be divided into four main classes of sensors:

- *thermal sensors*: the ΔH (enthalpy variation) of reaction with the sensitive material is used;
- *mass sensors*: piezoelectric microbalances to determine the accumulation;
- *electrochemical sensors*: detect changes in conductivity in transistors;

- *optical sensors*: fluorescence, chemiluminescence, etc. give greater complexity but strong selectivity.

Selectivity and reversibility are extremely important for the functioning of the sensor. To achieve this, it is required that the interaction between the analyte molecules and the chemically active material of the sensor is not too intense and that it is possibly specific. Another parameter that strongly affects the applicability range of a sensor is its mechanical, thermal, chemical robustness.

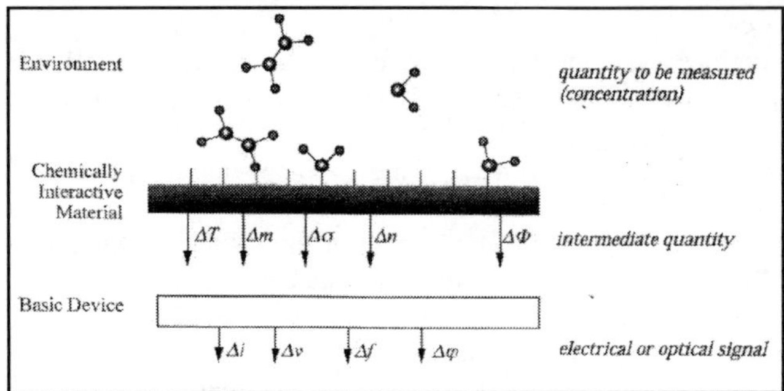

Figure 3. Operating scheme of a chemical-biological sensor.

Ideally, therefore, a chemical sensor is required to have:
- high sensitivity;
- wide dynamics;
- specific response for the analyte;
- rapid response and recovery times and long-term stability [19].

Recent Results by Theoretical Modelling

It has been performed a new Drude-Lorentz-type model, said DS model, based on the Fourier transform of the complex conductivity $\sigma(\omega)$; it gives the analytical expressions of the velocities correlation function $<\vec{v}(t)\cdot\vec{v}(0)>_T$, the mean squared deviation of position $R^2(t)$ and the diffusion coefficient D. It is available in the classical, quantum and relativistic versions; the global quantum-relativistic form is in progress [20-23].

The quantum expressions of D are:

$$D=2\left(\frac{KT}{m^*}\right)\sum_{i=0}^{n}\left(\left[\frac{f_i\,\tau_i}{\alpha_{iR}}\sin\left(\frac{\alpha_{iR}}{2}\frac{t}{\tau_i}\right)\exp\left(-\frac{t}{2\tau_i}\right)\right]\right) \tag{1}$$

$$D=\left(\frac{KT}{m^*}\right)\sum_{i=0}^{n}\left(\left(\frac{f_i\,\tau_i}{\alpha_{iI}}\right)\left(-\exp\left(-\frac{1+\alpha_{iI}}{2}\frac{t}{\tau_i}\right)+\exp\left(-\frac{1-\alpha_{iI}}{2}\frac{t}{\tau_i}\right)\right)\right) \tag{2}$$

The parameters α_{hK} of the model are two real numbers defined as follows:

$$\alpha_{iR}=\sqrt{4\tau_i^2\,\omega_i^2-1}\;;\;\alpha_{iR}\in\Re^+ \tag{3}$$

$$\alpha_{iI}=\sqrt{1-4\tau_i^2\,\omega_i^2}\;;\;\alpha_{iI}\in[0,1]\subset\Re \tag{4}$$

where τ_i and ω_i are relaxation times and frequencies of modes. $f_i=1$, $\tau_i=\tau$ and $\omega_i=\omega_0$ (center frequency) give the classical expressions; K is the Boltzmann's constant and m^* the effective mass. Two behaviours of D are therefore possible: a damped oscillation and/or the superposition of two diverse exponentials.

Variables influencing D and next the sensitivity are:

1) the system's temperature T;

2) the parameter $\alpha = \alpha(\tau_i, \omega_i)$;

3) the effective mass m^*, connected to the physical and chemical processing on materials, as the doping;

4) variations of the chiral vector [24,25];

5) the weights of each mode, being:

$$\omega_{p_i}^2 = \frac{4\pi N e^2}{m^*} f_i \tag{5}$$

6) the possible variation of the initial peak of D and its value during time via a tuning of the carriers velocity.

We show the behaviour of D for CNTs with change of T, with $m^* = 0.5 m_e$ (m_e = electron mass), $\tau = 0.17 \cdot 10^{-12} s$, $\alpha_I = 0.5$ [26-28] (Fig. 4).

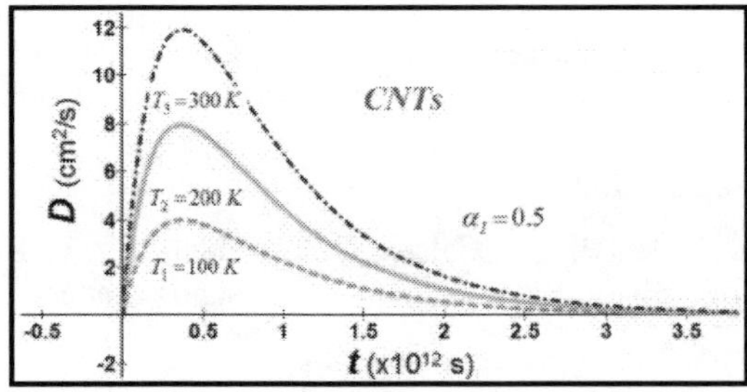

Figure 4. D vs t for CTNs, varying T.

We can compare values of diffusion considering different nanomaterials [29-31] (Fig. 5). Interesting results can be obtained considering variations of the chiral vector $C_h = (n, m)$, so as the variation of α, τ and of previously indicated variables [32-35].

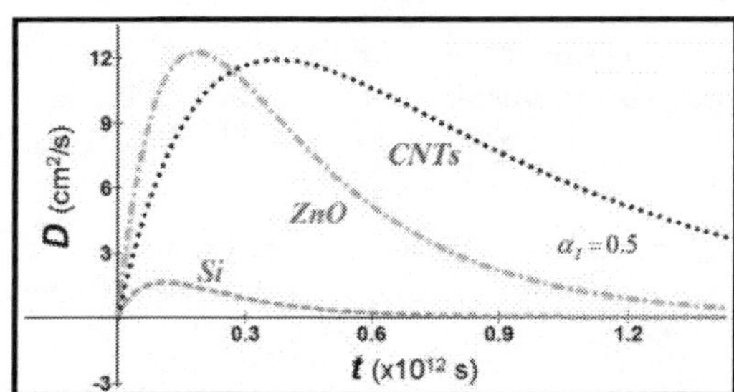

Figure 5. Diffusion in time for different nanomaterials.

Conclusions

The field of graphene is today rapidly evolving from pure science to technology. Different applications require graphene-based materials with different properties. The current and future market of graphene applications is driven by the production strategies of this material, allowing its widespread practical implementation.

The cheapest GRMs with less stringent conditions can appear on the market, used in flexible electronic instruments based on conductive inks, as flexible solar cells and batteries, and for equipments such as spin valves and non-volatile memories.

As in many other fields, research on biomedical applications of graphene is having rapid progress, and is rapidly expanding. The progress made in this field so far is exciting and encouraging, but with high challenges to be overcome. These goals can only be achieved by the joint efforts of chemistry, biomedicine, materials science, nanotechnology, physics, mathematical modelling.

References

[1] M.I. Katsnelson, Graphene: carbon in two dimensions, Materials Today 10(1-2) (2007) 20-27.

[2] K.S. Novoselov, A.K. Geim, S.V. Morozov, D. Jiang, Y. Zhang, S.V. Dubonos, I.V. Grigorieva, A.A. Firsov, Electric field effect in atomically thin carbon films, Science 306(5696) (2004) 666-669, doi: 10.1126/science.1102896.

[3] J.H. Chen, C. Jang, S. Xiao, M. Ishigami, M.S. Fuhrer, Intrinsic and extrinsic performance limits of graphene devices on SiO_2, Nature Nanotechnology 3(4) (2008) 206-9, doi: 10.1038/nnano.2008.58.

[4] K.S. Novoselov, S.V. Morozov, T.M.G. Mohinddin, L.A. Ponomarenko, D.C. Elias, R. Yang, I.I. Barbolina, P. Blake, T.J. Booth, D. Jiang, J. Giesbers, E.W. Hill, A.K. Geim, Electronic properties of graphene, Physica Status Solidi b 244(11) (2007) 4106-4111, https://doi.org/10.1002/pssb.200776208.

[5] A.A. Balandin, S. Ghosh, W. Bao, I. Calizo, D. Teweldebrhan, F. Miao, C.N. Lau, Superior Thermal Conductivity of Single-Layer Graphene, Nano Letters 8(3) (2008) 902-907, https://doi.org/10.1021/nl0731872.

[6] K.S. Novoselov, V.I. Fal'ko, L. Colombo, P.R. Gellert, M.G. Schwab & K. Kim, A roadmap for graphene, Nature 490 (2012) 192-200, doi:10.1038/nature11458.

[7] S. Bae, H. Kim, Y. Lee, X. Xu, J.-S. Park, Y. Zheng, J. Balakrishnan, T. Lei, H.R. Kim, Y.I. Song, Y.-J. Kim, K.S. Kim, B. Özyilmaz, J.-H. Ahn, B.H. Hong & S. Iijima, Roll-to-roll production of 30-inch graphene films for transparent electrodes, Nature Nanotechnology 5 (2010) 574-578, doi:10.1038/nnano.2010.132.

[8] T. Kobayashi, M. Bando, N. Kimura, K. Shimizu, K. Kadono, N. Umezu, K. Miyahara, S. Hayazaki, S. Nagai, Y. Mizuguchi, Y. Murakami, and D. Hobara, Production of a 100-m-long high-quality graphene transparent conductive film by roll-to-roll chemical vapor deposition and transfer process, Applied Physics Letters 102 (2013) 023112, https://doi.org/10.1063/1.4776707.

[9] K.R. Ratinac, W. Yang, S.P. Ringer, F. Braet, Toward Ubiquitous Environmental Gas Sensors-Capitalizing on the Promise of Graphene, Environmental Science & Technology 44(4) (2010) 1167-1176, https://doi.org/10.1021/es902659d.

[10] S.A. Naghdehforooshha, G. Moradi, Plasmonic wave propagation mode analysis of single and multi-layer graphene-pec structures, Optik 200 (2020) 163365, https://doi.org/10.1016/j.ijleo.2019.163365.

[11] F. Schedin, A.K. Geim, S.V. Morozov, E.W. Hill, P. Blake, M.I. Katsnelson & K.S. Novoselov, Detection of individual gas molecules adsorbed on graphene, Nature Materials 6 (2007) 652-655, doi:10.1038/nmat1967.

[12] P. Dutta and P.M. Horn, Low-frequency fluctuations in solids: 1/f noise, Reviews of Modern Physics 53 (1981) 497.

[13] C.O. Park, J.W. Fergus, N. Miura, J. Park, A. Choi, Solid-state electrochemical gas sensors, Ionics 15(3) (2009) 261-284, https://doi.org/10.1007/s11581-008-0300-6.

[14] X. Chen, B. Chen, Macroscopic and Spectroscopic Investigations of the Adsorption of Nitroaromatic Compounds on Graphene Oxide, Reduced Graphene Oxide, and Graphene Nanosheets, Environmental Science & Technology 49(10) (2015) 6181-6189, https://doi.org/10.1021/es5054946.

[15] J. Wei, Z. Zang, Y. Zhang, M. Wang, J. Du, and X. Tang, Enhanced performance of light-controlled conductive switching in hybrid cuprous oxide/reduced graphene oxide (Cu$_2$O/rGO) nanocomposites, Optics Letters 42(5) (2017) 911-914, https://doi.org/10.1364/OL.42.000911.

[16] P. Di Sia, About the Influence of Temperature in Single-Walled Carbon Nanotubes: Details from a new Drude-Lorentz-like Model, Applied Surface Science 275 (2013) 384-388.

[17] Z. Sun, T. Hasan, F. Torrisi, D. Popa, G. Privitera, F. Wang, F. Bonaccorso, D.M. Basko, A.C. Ferrari, Graphene Mode-Locked Ultrafast Laser, ACS Nano 4(2) (2010) 803-810, https://doi.org/10.1021/nn901703e.

[18] P. Di Sia, Present and Future of Nanotechnologies: Peculiarities, Phenomenology, Theoretical Modelling, Perspectives, Reviews in Theoretical Science 2(2) (2014) 146-180, https://doi.org/10.1166/rits.2014.1019.

[19] P. Di Sia, An Analytical Transport Model for Nanomaterials, Journal of Computational and Theoretical Nanoscience 8 (2011) 84-89.

[20] P. Di Sia, An Analytical Transport Model for Nanomaterials: The Quantum Version, Journal of Computational and Theoretical Nanoscience 9(1) (2012) 31-34.

[21] P. Di Sia, Relativistic nano-transport and artificial neural networks: details by a new analytical model, International Journal of Artificial Intelligence and Mechatronics (IJAIM) 3(3) (2014) 96-100.

[22] P. Di Sia, A new analytical transport model for (nano)physics, International Research Journal of Engineering and Technology (IRJET) 2(7) (2015) 1-4.

[23] P. Di Sia, Quantum-Relativistic Velocities in Nano-Transport, Applied Surface Science 446 (2018) 187-190, https://doi.org/10.1016/j.apsusc.2018.01.273.

[24] J.M. Marulanda, A. Srivastava, Carrier Density and Effective Mass Calculation for carbon Nanotubes, Physica Status Solidi (b) 245(11) (2008) 2558-2562.

[25] P.G. Collins, A. Zettl, H. Bando, A. Thess, and R.E. Smalley, Nanotube Nanodevice, Science 278(5335) (1997) 100-102.

[26] H. Altan, F. Huang, J.F. Federici, A. Lan, and H. Grebel, Optical and electronic characteristics of single walled carbon nanotubes and silicon nanoclusters by tetrahertz spectroscopy, Journal of Applied Physice 96 (2004) 6685-6689.

[27] I. Pirozhenko, A. Lambrecht, Influence of slab thickness on the Casimir force, Physical Review A 77 (2008) 013811-013818.

[28] J.B. Baxter, C.A. Schmuttenmaer, Conductivity of ZnO Nanowires, Nanoparticles, and Thin Films Using Time-Resolved Terahertz Spectroscopy, Journal of Physical Chemistry B 110 (2006) 25229-25239.

[29] P. Parkinson, H.J. Joyce, Q. Gao, H.H. Tan, X. Zhang, J. Zou, C. Jagadish, L.M. Herz, and M.B. Johnston, Carrier Lifetime and Mobility Enhancement in Nearly Defect-Free Core-Shell Nanowires Measured Using Time-Resolved Terahertz Spectroscopy, Nano Letters 9(9) (2009) 3349-3353.

[30] C. Soldano, A. Mahmood, E. Dujardin, Production, properties and potential of graphene, Carbon 48(8) (2010) 2127-2150, https://doi.org/10.1016/j.carbon.2010.01.058.

[31] M.A. Rafiee, J. Rafiee, I. Srivastava, Z. Wang, H. Song, Z.-Z. Yu, N. Koratkar, Fracture and Fatigue in Graphene Nanocomposites, Small 6(2) (2010) 179-183.

[32] Y. Shao, J. Wang, H. Wu, J. Liu, I.A. Aksay, Y. Lin, Graphene Based Electrochemical Sensors and Biosensors: A Review, Electroanalysis 22(10) (2010) 1027-1036.

[33] J.T. Robinson, F.K. Perkins, E.S. Snow, Z. Wei, P.E. Sheehan, Reduced Graphene Oxide Molecular Sensors, Nano Letters 8(10) (2008) 3137-3140.

[34] M. Pumera, A. Ambrosi, A. Bonanni, E.L.K. Chng, H.L. Poh, Graphene for electrochemical sensing and biosensing, Trends in Analytical Chemistry 29(9) (2010) 954-965, doi: 10.1016/j.trac.2010.05.011.

[35] P. Di Sia, Mathematics and Physics for Nanotechnology: Technical Tools and Modelling, first ed., CRC Press - Taylor & Francis Group, Jenny Stanford Publishing, Singapore, 2019.

Nano Hybrids and Composites
ISSN: 2297-3370, Vol. 33, pp 73-81
© 2021 Trans Tech Publications Ltd, Switzerland

Submitted: 2020-10-13
Revised: 2021-08-01
Accepted: 2021-08-13
Online: 2021-10-11

Study of the Effect of Physical Aging on the Polylactic Acid by Thermal and Optical Techniques

Amirouche BOUAMER[1,2,a*], Nasser BENREKAA[1,b]
and Abderrahmane YOUNES[2,c]

[1]Materials Physics Laboratory, Department of Physics, University of Science and Technology Houari Boumediene (USTHB), Algiers, Algeria

[2]Research Center in Industrial Technologies (CRTI), P.O.Box 64,Cheraga 16014 Algiers, Algeria

[a]amirouchebouamer@gmail.com, [b]nbenrekaa@gmail.com, [c]younesabdo11@gmail.com

Keywords: Polylactic acid, Physical aging, Glass transition temperature, Differential scanning calorimetry, Thermally Stimulated Depolarization Current, Attenuated total reflection.

Abstract: In this work, we investigated the influence of physical aging on polylactic acid (PLA) films using thermal and optical techniques; Differential Scanning Calorimetry (DSC), Thermally Stimulated Depolarization Current (TSDC), and Attenuated Total Reflection Spectroscopy (ATR). The PLA films were aged for different periods: 60, 90, and 120 minutes at a temperature $T_a = 43$ °C. The result obtained by DSC showed that the effect of physical aging appeared as an endothermic peak, which increased with increasing aging time and evolved towards higher temperatures. TSDC results showed a thermal current peak located between 30 and 80 ° C, which represented the main relaxation mode (α relaxation) of the dielectric manifestation of the glass transition. The intensity of this peak decreased and was shifted to higher temperatures when aging time increased, this result can be explained by a decrease in the molecular mobility of macromolecular chains due to the decrease in the free volume. The effect of physical aging on the PLA by the ATR technique showed a gradual decrease in all absorption bands during the aging period. In particular, the wide absorption band between 3000 and 3700 cm^{-1} attributed to the hydroxyl group (OH), which disappeared after two hours of aging.

1. Introduction

In the recent years, the use of the polylactic acid (PLA) augmented significantly because of its importance in various areas such as the industrial engineering fabrication and the field of food packaging due to its availability and competitive global properties [1-4]. PLA is a biodegradable amorphous or semi-crystalline material issued from renewable resource. Its crystallinity is low and can be changed by plasticization, combination or reactive treatment [5]. PLA has a low glass transition temperature and these properties change with the physical aging time [6-8]. The physical aging of polymers is generally known through changes in molecular motions that are far from their random arrangement due to the non-equilibrium thermodynamic state at temperatures below the glass transition temperature (Tg). As a result, the conformation of the molecules changes quietly through the use in order to reach the equilibrium state. Several factors influence the rate of this change, such as the chemical structure and the thermal history of the polymer. Physical aging is generally assumed to go ahead to the augmented Tg and decreased free volume [9-11]. Many researchers have focused on the study of the physical aging of PLA, given its importance in all fields; industrial and medical. Most of these studies were carried out using thermal analysis for calorimetric measurements, such as Differential Scanning Calorimetry (DSC). During Physical aging, samples are generally annealed at temperatures below glass transition temperature and then heated in order to determine the transition enthalpy and Tg. The aging temperature changes the characteristics cited above, but we have several factors, which influence the glass transition temperature, such as the aging time. The combined use of thermal and optical techniques in the study of the effect of physical aging on polymers is much less reported in published works. The use of polymers, particularly PLA, requires knowledge of their physicochemical properties, in order to optimize the fields of use. Determining the useful period of PLA in medical device is of great importance. The combination of the DSC measures, Thermally Stimulated Depolarization currents (TSDC), and attenuated total reflectance (ATR) measurements represent a powerful experimental means used to study the effect of physical aging of polymers. The purpose of this work is to study the influence of the physical aging time on the polylactic acid (PLA) by thermal analysis and Thermally Stimulated Depolarization current (TSDC) techniques, and the Attenuated Total Reflectance (ATR) measurements were used to highlight structural changes in PLA under different Physical aging time; the results are discussed in comparison with the works published in the literature.

2. Experimental

2.1. Differential Scanning Calorimeter

The PLA films used in the present work were provided by NatureWorks. The procedure of physical aging and the measurements of the glass transition temperature and excess of enthalpy relaxation were carried out using a DSC7 Perkin-Elmer apparatus. The temperature of the instrument was calibrated with indium and lead. The calibration of the heat flow was provided by the standards of even indium sample. The recording of the DSC transition spectra was carried out during heating with a heating rate$10°C/min$. The samples, encapsulated in aluminum cups (mass $\sim$ 10 mg), were first heated to 200 °C and remained there isothermally during 5 minutes to eliminate the previous thermal history of the samples, then the specimens were rapidly cooled to obtain amorphous samples, and put in an oven at an aging temperature $T_a = 43 \, °C$ for different periods of aging from 60 to 120 minutes. Specimens were then reheated to 200°C for the measurement of enthalpy relaxation and glass transition.

2.2. Thermally Stimulated Depolarization Current (TSDC)

The TSDC measurements were performed with a device that has been realized within our lab (dielectric) according to the concept described by Bucci Fieschi [12]. This system is composed of three compartments: A measurement cell, which consisted of a hollow cylinder that contained two horizontal circular electrodes. The latter formed a plane capacitor, in the middle of which the sample to be studied is placed. A pumping unit consisted of a primary pump, Alcatel-type paddles coupled with an oil diffusion secondary pump. The polarization of the specimen was carried out by means of a voltage source HP712C over the range (0 - 600) volts. An electronic measurement system, in which the reading of temperature was done by a digital reader AOIP analog output, was used. The latter was connected to the platinum probe PT100. The thermo-stimulated currents were measured using Keithley electrometer 610 C of 10^{-16} Ampere limit sensitivity. The principle of this technique is as follows: The sample is positioned between two metal electrodes of a condenser placed in a measurement cell. The latter is maintained at a constant pressure of 400 Torr under nitrogen gas. The sample is polarized at the polarization temperature T_p, under an electric field E_p for a period of time Δt_p longer than the polarization relaxation time, in order to reach saturation. The sample is cooled down, under E_p to a temperature T_0. The E_P is then removed, the sample is quenched to the liquid nitrogen temperature and the electrodes are short-circuited in order to eliminate the residual surface charge. The capacitor is then connected to the input of a sensitive electrometer Keithley 610C. The temperature is measured using a Pt100 probe placed close to the sample. Finally, the depolarization current is measured during reheating of the sample at constant rate b by an electrometer connected to a computer for the data acquisition.

2.3. Attenuated total reflection (ATR) Spectroscopy

The attenuated total reflection spectroscopy, allowed us, to detect new bonds in the Polymer and to observe the influence of heat treatment (physical aging) on previously existing bonds through the variation of their frequency and intensity. The samples were analyzed by using an (Agilent CARY 360 SPECTROMETER, Germany). All samples spectra were obtained in absorbance mode with a minimum of 32 scans and an average signal resolution of 4 cm^{-1} in a spectral range of 4000–400 cm^{-1}.

3. Results and Discussion

3.1. Thermal analysis by DSC.

The DSC curves of PLA films annealed at T_a= 43°C for different aging periods are shown in Figure 1. The appearance of an endothermic peak at the glass transition region shows that the structure of PLA changed under the effect of physical aging. The intensity of this peak increased with increasing aging time. This phenomenon can be explained by considering the reduction in free volume due to the structural relaxation of the material [13]; the aged sample has a smaller free volume as well as a lower enthalpy and potential energy than the unaged sample. Therefore, during heating in the DSC test, more energy was required for the glass transition, resulting in an increase in the area below the endothermic peak [14].

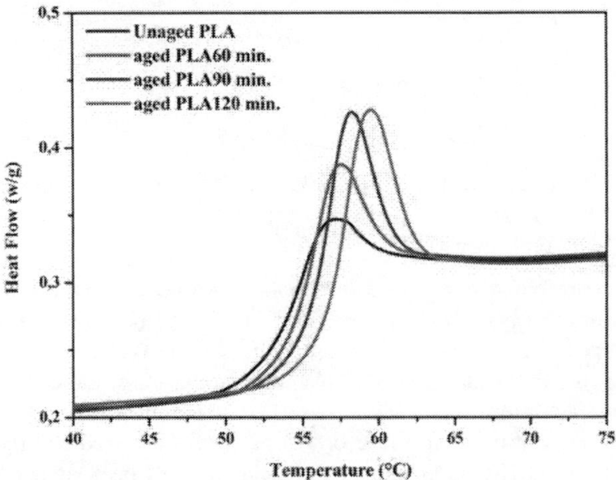

Figure 1 DSC curves in the glass –transition region of polylactic acid annealed for different aging time t_a at T_a=43°C, heating rate 10 °C/min

Figure 2 shows an increase of the excess enthalpy of relaxation (ΔH) with increasing aging time t_a. This can be explained by considering the macromolecular chains of the polymer that try to reach the equilibrium energy of weak conformations. The glass transition temperature (Tg) increased with increasing aging time, due to the molecular arrangement in the polymer which evolved towards an equilibrium state. Consequently, the free volume is reduced, which decreased the molecular mobility in the glassy state [15].

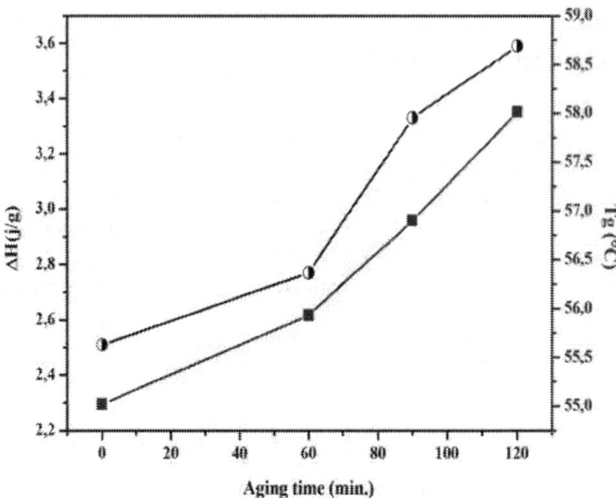

Figure 2 Evolution of the Excess of enthalpy relaxation (ΔH) and glass transition (T_g) as a function of aging time t_a

3.2. Studies by the TSDC technique.

The global spectrum represented in figure 3 was obtained after polarization of the sample for two minutes at the temperature of polarization $T_p = 62\ °C$ under an polarization electric field $E_p = 1,\ 4 \times 10^6\ V.m^{-1}$. Once the polarization time has elapsed, the sample was cooled to the temperature $T_o = -30\ °C$ with a cooling rate $q = -10\ °C\ /\ min$.   At this temperature, the electric field was removed and the sample had been short-circuited for a period $\Delta t = 2$ min, in order to eliminate the residual currents. Raising the temperature at the rate $b = 7\ °C\ /\ min$ allowed the recording of the global spectrum shown in figure 3. The latter shows a thermo current peak at temperatures around 30 - 80 °C, and its maximum peak located at a temperature $T_M = 62\ °C$, which represents the α relaxation mode related to the dielectric manifestation of the glass transition temperature [16, 17].

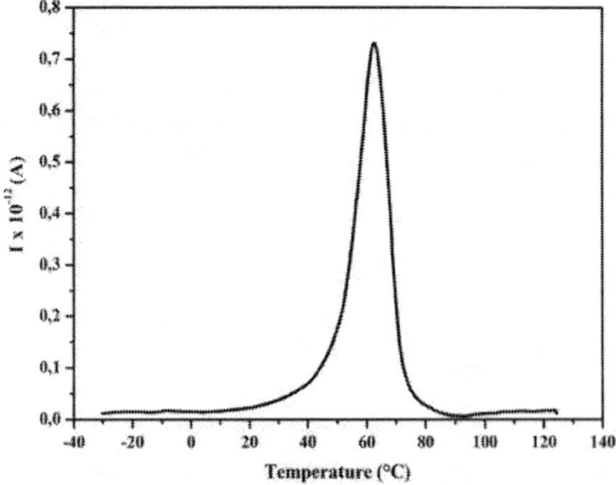

Figure 3 Global spectrum TSDC of PLA poled at $T_p = 62\ °C$ and $E_p = 1.6$ MV/m. as a function of temperature

Figure 4 shows TSDC curves of PLA films, which illustrate a superposition of thermo-current spectra, obtained after annealing of the samples at a temperature $T_o = 43\ °\ C$ for various periods ranging from 60 to 120 minutes with an interval of 30 min. We observe in figure 4 a significant decrease in the intensity of the thermo-current peaks, which are shifted to high temperatures with increasing aging time. This changes correspond to the effect of physical ageing time on the main relaxation mode (α relaxation), as reported in literature [16].

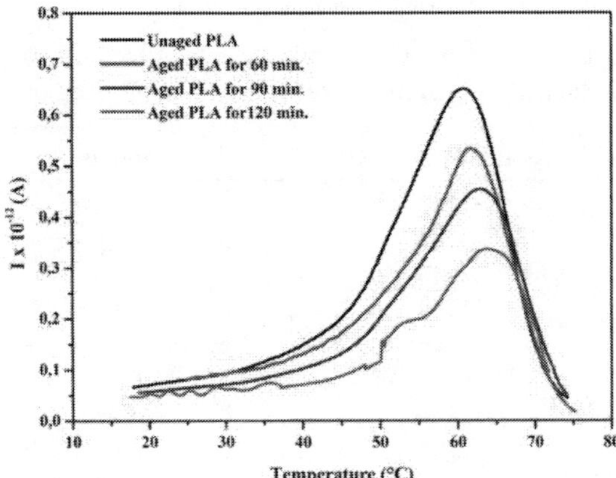

Figure 4 TSDC spectra of PLA films aged at $T_a = 43$ °C for different durations

Effect of the physical aging time on the α relaxation mode is depicted in Figure 4. It is obviously noticed that an increase in the aging time induced a shift of the thermo-stimulated currents peaks towards higher temperatures. The same effect was observed in the result obtained by the DSC analysis shown in figure 1. The diminution of the thermo-currents peak intensity and polarization P_{max} can be attributed to the reduction of the free volume, which leaded to the decrease of the molecular mobility. This decrease made difficult the dipoles movements. Therefore, their orientation according to the electric field direction would be weaker. Hence, a less important polarization than the one corresponding to the non-aged material was recorded. The diminution of the activation enthalpies shown in figure 5 and calculated by the half-height method could be explained by the fact that the reduction of the free volume leaded to a less important disorder of the macromolecular chains within the material. Consequently, a low activation entropy will be recorded, which resulted in a reduction of the activation enthalpy. These findings agree well with those reported in literature [18].

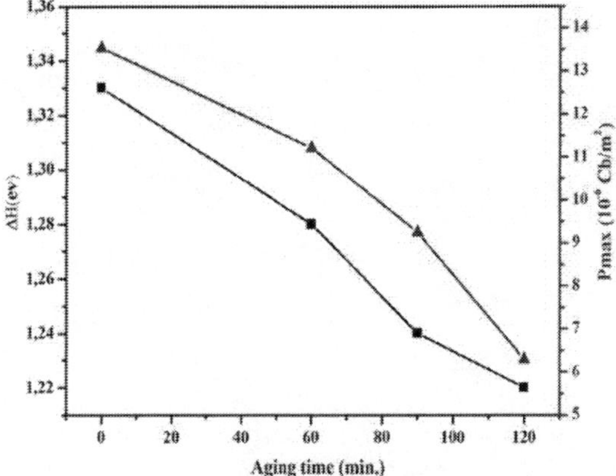

Figure 5 Evolution of the saturation polarization and Excess enthalpy relaxation ΔH as a function of aging time t_a

3.3. Studies by Attenuated Total Reflectance (ATR):

The FTIR-ATR spectra of unaged and aged PLA films during various periods: 60, 90 and 120 minutes are displayed in Figure 6. The vibrational assignments of absorbance Spectra FTIR- ATR of the pure and aged PLA films are summarized in the table 1.

Table 1 Wave numbers (cm^{-1}) and vibrational assignments of (ATR) absorbance Spectra of PLA and aged PLA films in the 4000–600 cm^{-1} Region.

Wave numbers [cm^{-1}]	Assignments	
3375	ν	OH
2994	ν_{as}	CH3
2944	ν_{s}	CH3
2843	ν_{s}	CH
1746	ν	C=O
1451	δ_{as}	CH3
1358	δ_{s}	CH3
1265	$\delta+\nu$	CH + COC
1179	ν_{as}	COC+CH3
1125	ν_{as}	CH3
1080	ν	COC
1040	ν	C-CH3
956	ν	CH3+ CC
866	ν	C-COO

The absorptions bands at 2994 cm^{-1} and 2944 cm^{-1} are attributed to the asymmetric and symmetric C-H stretching vibrations of the -CH$_3$ group respectively [19]. The band centered at 1746 cm^{-1} is assigned to the C=O stretching of the ester group of PLA [20-22]. The bending frequencies for asymmetric and symmetric -CH$_3$ group were identified at 1451 and 1358 cm^{-1}, respectively. The absorption peak for asymmetric C-O-C stretching vibration was observed at 1080 cm^{-1}. The band at 956 cm^{-1} was assigned to the characteristic vibration of the helical backbone (C-C) with CH$_3$ rocking mode [22]. Two bands related to the PLA crystalline and amorphous phases were found at 867 and 754 cm^{-1} respectively [23]. The PLA characteristic peaks showed in figure 6 are in agreement with those found in literature [24].

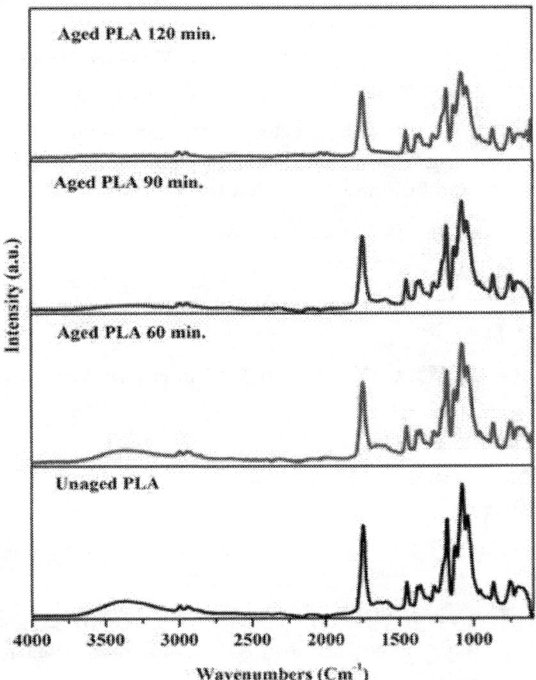

Figure 6 ATR spectra of PLA and aging PLA for different periods: 0min, 60min, 90min and 120min

An increase in aging time affected all the absorption bands of PLA, particularly the broad band located between 3700–3000 cm^{-1}, which characterizes the hydroxyl group (–OH) [25]. The hydroxyl group band decreased significantly with increasing aging time when compared to the unaged and aged films. This is consistent with the evolution of the main relaxation mode (α relaxation) obtained previously by the Thermally Stimulated Depolarization Current study. Our result agrees well with the one found by Moudoud et al. [26].

4. Conclusions

The effect of physical aging time on PLA has been studied by the DSC technique. The phenomenon appeared in the form of an endothermic peak, which increased with increasing aging time and moved towards higher temperatures. The excess of enthalpy relaxation and the glass transition temperature also increased with increasing aging time; this behavior is explained by a reduction in free volume and molecular mobility due to the structural relaxation of the polylactic acid.

The technique of thermally stimulated depolarization current was used to study the effect of physical aging on the relaxation mode near the glass transition temperature. It showed a decrease of the main relaxation mode (α relaxation) intensity during aging due to a decrease in molecular mobility in the material.

The reduction of α relaxation mode has been found by the optical technique (ATR): it was highlighted by a decrease in all absorption bands intensities due to the reduction of the free volume and molecular mobility during aging.

References

[1]   F. Galiano, K. Briceno, T. Marino, A. Molino, K.V. Christensen, A. Figoli, Advances in biopolymer-based membrane preparation and applications, J. Memb. Sci. 564(2018) 562-586.

[2]   B. Bhushan, R. Kumar, Plasma treated and untreated thermoplastic biopolymers/biocomposites in tissue engineering and biodegradable implants, Materials for Biomedical Engineering, Hydrogels and Polymer-Based Scaffolds, 2019,pp. 339-369

[3]   K. Jin, Y.j. Tang, X. Zhu, Y. Zhou, International Journal of Biological Macromolecules, 1621 (2020) 1109-1117

[4]   R. Auras, B. Harte, S. Selke, An overview of polylactides as packaging materials, Macromol. Biosci. 4 (2004) 835-864

[5]   E.T.H. Vink, K.R. Rabago, D.A. Glassner, P.R. Gruber, Applications of life cycle assessment to NatureWorks™ polylactide (PLA) production, Polym. Degrad. Stab. 80 (2003) 403-419

[6]   A. Celli, M. Scandola, Thermal properties and physical ageing of poly (l-lactic acid) Polymer 33 (1992)2699 - 2703

[7]   S. Kawana, R.A.L. Jones, Effect of physical ageing in thin glassy polymer films, Eur. Phys. J. E. 10 (2003) 223-230

[8]   M.C. Righetti, N. Delpouve, A. Saiter, Physical ageing of semi-crystalline PLLA: Role of the differently constrained amorphous fractions , AIP. Conf. Proc. 1981 p 20-82 (2018)

[9]   G. M. Odegard, A. Bandyopadhyay, Physical Aging of Epoxy Polymers and Their Composites, J. of Poly. Science PART B: Polymer Physics, 49   (2011)1695–1716

[10] JH. Guo, RE. Robertson, and GL. Amidon, Influence of physical Aging on Mechanical Properties of polymer Free Films:The prediction of Long-Term Aging effects on the Water Permeability and dissolution Rate of Polymer Film-Coeted Tablets Pharmaceutical Research, 8, 12, (1991) 1500-1504

[11] G. M. Nasr, M. Amin and H. M. Shaker, Influence of Physical Aging on the Dielectric Relaxation of Ethylene-co-Vinyl Acetate (EVA)/PolyPyrrole (PPy) Interpenetrating Polymer Network Composite Using TSDC Technique Journal of the Korean Physical Society, 58, 2, (2011)301-305

[12] C. Bucci, R. Fieschi, G. Guidi, Ionic Thermocurrents in Dielectrics, Physical Review (Series I), 148, 2 (1966)

[13] K.liao, D. Quan, Z.Lu, Effects of physical aging on glass transition behavior of poly (DL-lactide), European Polymer Journal, 38, (2002) 157-162

[14] P. Pan, B. Zhu, Y. Inoue, Enthalpy Relaxation and Embrittlement of Poly(l-lactide) during Physical Aging, Macromolecules, 40, 26 (2007) 9664-9671

[15] N.M. Alves,J.F. Mano,E. Balaguer,J.M. Meseguer Dueñas, and J.L. Gómez, Glass transition and structural relaxation in semi-crystalline poly (ethylene terephthalate): A DSC study, Polymer, 43 (2002) 4111- 4122

[16] J.A. Diego, J.C. Canadas, M. Mudarra, J. Belana, Glass transition studies in physically aged partially crystalline poly (ethylene terephthalate) by TSC, Polymer, 40, (1999) 5355-5363

[17] N. Benrekaa, A. Gourari, M. Bendaoud, and K. Ait-hamouda, Analysis of thermally stimulated current and effect of rubbery annealing around glass-rubber transition temperature in polyethylene terephthalate, Thermochim. Acta 413 (2004) 39 – 46

[18] M.D.Joaquim, J.M. RamosNatália, T. CorreiaNatália, T. Correia, Ageing exploration of the energy landscape of a glass by the TSDC technique, Molecular Physics, 100(16) (2002) 2669-2677

[19] M.Savaris, GL. Braga, V.dos Santos, GA. Carvalho, A. Falavigna, DC. Machado, C. Viezzer, RN. Brandalise, Biocompatibility Assessment of Poly(lactic acid) Films after Sterilization with Ethylene Oxide in Histological Study In Vivo with Wistar Rats and Cellular Adhesion of Fibroblasts In Vitro, International Journal of Polymer Science, vol. 2017, Article ID 7158650 (2017)

[20] P. Qu, Y. Goa, GF. Wu and LP. Zhang, Nanocomposites of Poly(Lactic Acid) reinforced with Cellulose Nanofibrils J. Bio Resources, 5, 3 (2010) 1811-1823

[21] X. Wen, Y. Lin, C. Han, K. Zhang, X. Ran, Y. Li, L. Dong, Thermomechanical and optical properties of biodegradable poly (L-lactide)/silica nanocomposites by melt compounding, journal of Applied Polymer Science, 114, 6 (2009) 3379-3388

[22] A. Bouamer, N. Benrekaa, A. Younes and H. Amar, Characterization of the Polylactic acid stretched uniaxial and annealed by Raman spectrometry and Differential scanning calorimetry, IOP Conf. Ser.: Mater. Sci. Eng.461 012006 (2018)

[23] E. Elisabeta, M. Râpă, O. Popa, M. Elena, P. Mona, Polylactic Acid/Cellulose Fibres Based Composites for Food Packaging Applications Materiale Plastice, 4 (2017) 673-677

[24] C.C. Chen, J.Y. Chueh, H. Tseng, H.M. Huang, and S.Y. Lee, Preparation and characterization of biodegradable PLA polymeric blends, Biomaterials, 24, 7 (2003) 1167-1173

[25] P. Denis, M. Wrzecionek, AG.Gajadhur and P.Sajkiewicz, Poly (Glycerol Sebacate)-Poly(l-Lactide) Nonwovens. Towards Attractive Electrospun Material for Tissue Engineering, Polymers, 11, 2113 (2019) 1-26

[26] M. Moudoud, A. Hedir, O. Lamrous, S. Diaham and T. Touam, Physical ageing of insulating polystyrene from dielectric properties measurements and structural analysis, Mater. Res. Express, 6, 095324 (2019)

Nano Hybrids and Composites
ISSN: 2297-3370, Vol. 33, pp 83-92
© 2021 Trans Tech Publications Ltd, Switzerland

Submitted: 2021-02-27
Revised: 2021-05-13
Accepted: 2021-05-19
Online: 2021-10-11

Dynamic Behavior Study of Functionally Graded Porous Nanoplates

Hadj Mostefa Adda[1,a], BOUCHAFA Ali[2,b*], MERDACI Slimane[3,c*]

[1]Industrial Engineering and Sustainable Development Laboratory, University of Rélizane, Instituteof Science & Technology, Civil Engineering Department, Algeria.

[2]University of Sidi Bel Abbes, Faculty of Technology, Civil Engineering Department, Algeria.

[3]Structures and Advanced Materials in Civil Engineering and Public Works Laboratory, University of Sidi Bel Abbes, Faculty of Technology, Civil Engineering and Public Works Department, Algeria.

[a]hadj9510@gmail.com, [a]addahadjmostefa@yahoo.fr, [b]bouchafa2006@yahoo.fr, [c]*slimanem2016@gmail.com

Keywords: FG-nanoplate; Non-local theory; Scale effect; Dynamic; Porosity.

Abstract. This paper introduces the analytical solutions of complex behavior analysis utilizing high-order shear deformation plate theory of functionally graded FGM nano-plate content consisting of a mixture of metal and ceramics with porosity. To incorporate the small-scale effect, the non-local principle of elasticity is used. The impact of variance of material properties such as thickness-length ratio, aspect ratio, power-law exponent and porosity factor on natural frequencies of FG nano-plate is examined. Compared to those achieved from other researchers, the latest solutions are. Using the simulated displacements theory, equilibrium equations are obtained. Current solutions of the dimensionless frequency are compared with those of the finite element method. The effect of geometry, material variations of nonlocal FG nano-plates and the porosity factor on their natural frequencies are investigated in this review. The results are in good agreement with those of the literature.

Introduction

Functionally graded materials (FGM) are a new type of composite systems that are of particular interest in the design, usage, and manufacture of engineering. The material properties to be graded continuously across the thickness and to avoid sudden shifts in stress and displacement distributions for functionally graded structures. At each interface, the product is chosen according to specific applications and environmental loads. These materials have multiple advantages which can make them attractive from the point of view of their application potential. In addition to having a property gradation, it can be an improvement in rigidity, fatigue resistance, corrosion resistance or thermal conductivity, thus making it possible to increase or modulate performance. Such as reducing local constraints or even improving heat transfer [1-3]. The principle of functionally graded material was then considered in Japan in 1984 during a space plane project [4].

During the sintering process in the manufacture of FGMs, micro-porosities or voids can occur in the materials, due to the large difference in solidification temperature between the material constituents [5]. Porosities that are formed by a sequential multi-step infiltration technique inside FGM specimens [6]. As a consequence, more and more focus has been provided in recent years to studies dedicated to understanding the dynamic and free vibrations of the plate made of functionally graded FGM materials [7-14].

The investigated behavior of the nano-plate made of functionally graded materials based on small-scale effects by [15-17]. The non-local theory [18-19] is one of the well-known theories of continuum mechanics that involves small-scale effects with good precision. In order to investigate the static and dynamic characteristics of nano-plates [20-21], the various non-local theories, the differential form of elasticity of the Eringen [18] have received considerable attention. The analytical solutions of free vibration analysis of functionally graded nano-plates porous using non-local high order shear deformation plate theory presented in Ref [22]. For the study of the size-dependent free flexural vibration function of functionally graded nano-plates [23], the isogeometric centered finite element approach is used. For the model for static and vibration of dynamically graded micro/nano plates developed by Ref [24], updated couple stress and three-dimensional elasticity theories are used. Isotropic rectangular nano-plate bending and vibration utilizing non-local hypotheses and the principle of the third-order shear deformation plate [25]. The principle of nonlocal elasticity using the system of finite elements for the study of free vibration of nano-plates consisting of dynamically graded materials [26].

The intention of this paper is to examine the free vibration of non-local functionally graded nano-plates with porosities using high-order non-local shear deformation plate theory. Current solutions of the dimensionless frequency are compared with those of the finite element method presented by Zargaripoor et al [26] and the solutions obtained by other researchers. The consequences on the free vibration with nano-plates FG of various parameters are all discussed. Small effects are introduced using non-local elasticity theory. Through contrasting the present results with various theories, the validity of the theory is shown.

Theoretical and Mathematical Model

In the characteristic equations of this theory, the unorthodox contemporary ideas that the results of small scales were used by the nonlocal elasticity theory proposed by Eringen [18] are added. In nonlocal theory, through an integral equation, the stress tensor at point x of a physical environment is related to the strain tensor in the whole environment. A distinctive shape of the non-local constitutive relationship was suggested by Eringen as:

$$\sigma_x - \eta \frac{d^2 \sigma_x}{dx^2} = E\varepsilon_x \tag{1}$$

$$\tau_{xz} - \eta \frac{d^2 \tau_{xz}}{dx^2} = G\gamma_{xz} \tag{2}$$

Where the Young's and shear moduli, respectively, are E and G and the shear strain is γ.
η : is the non-local parameter represents the small-scale effect ($\eta = (e_0 a)^2$)
e_0: is a constant material.
a: is an internal characteristic length.

The nonlocal constitutive equations of an FGM nonlocal plate can be written as:

$$\left(1 - \eta\left(\frac{\partial^2}{\partial x^2} + \frac{\partial^2}{\partial y^2}\right)\right)
\begin{Bmatrix} N_x \\ N_y \\ N_{xy} \\ M_x^b \\ M_y^b \\ M_{xy}^b \\ M_x^s \\ M_y^s \\ M_{xy}^s \end{Bmatrix}
=
\begin{bmatrix}
A_{11} & A_{12} & 0 & B_{11} & B_{12} & 0 & B_{11}^s & B_{12}^s & 0 \\
A_{12} & A_{22} & 0 & B_{12} & B_{22} & 0 & B_{12}^s & B_{22}^s & 0 \\
0 & 0 & A_{66} & 0 & 0 & B_{66} & 0 & 0 & B_{66}^s \\
A_{11} & A_{12} & 0 & D_{11} & D_{12} & 0 & D_{11}^s & D_{12}^s & 0 \\
A_{12} & A_{22} & 0 & D_{12} & D_{22} & 0 & D_{12}^s & D_{22}^s & 0 \\
0 & 0 & A_{66} & 0 & 0 & D_{66} & 0 & 0 & D_{66}^s \\
B_{11}^s & B_{12}^s & 0 & D_{11}^s & D_{12}^s & 0 & H_{11}^s & H_{12}^s & 0 \\
B_{12}^s & B_{22}^s & 0 & D_{12}^s & D_{22}^s & 0 & H_{12}^s & H_{22}^s & 0 \\
0 & 0 & B_{66}^s & 0 & 0 & D_{66}^s & 0 & 0 & H_{66}^s
\end{bmatrix}
x
\begin{Bmatrix} \varepsilon_x^0 \\ \varepsilon_y^0 \\ \gamma_{xy}^0 \\ k_x^b \\ k_y^b \\ k_{xy}^b \\ k_x^s \\ k_y^s \\ k_{xy}^s \end{Bmatrix} \tag{3}$$

$$\left(1 - \eta\left(\frac{\partial^2}{\partial x^2} + \frac{\partial^2}{\partial y^2}\right)\right)
\begin{Bmatrix} S_{xz}^s \\ S_{yz}^s \end{Bmatrix}
=
\begin{bmatrix} A_{44}^s & 0 \\ 0 & A_{55}^s \end{bmatrix}
\begin{Bmatrix} \gamma_{xz} \\ \gamma_{yz} \end{Bmatrix} \tag{4}$$

Where the plate stiffness is determined by A_{ij}, B_{ij}, etc.

$$\left\{ A_{ij}, B_{ij}, D_{ij}, B^s_{ij}, D^s_{ij}, H^s_{ij} \right\} = \int\limits_{-h/2}^{h/2} \left(1, z, z^2, f(z), z f(z), f^2(z) \right) Q_{ij} dz \quad (i, j = 1, 2, 6)$$

$$\left\{ A^s_{ij} \right\} = \int\limits_{-h/2}^{h/2} \left([g(z)]^2 \right) Q_{ij} dz, \quad (i, j = 4, 5)$$

$$(5)$$

Consider a dense rectangular nano-plate FG made of functionally graded content with length a, width b and thickness h, as shown in Fig.1.

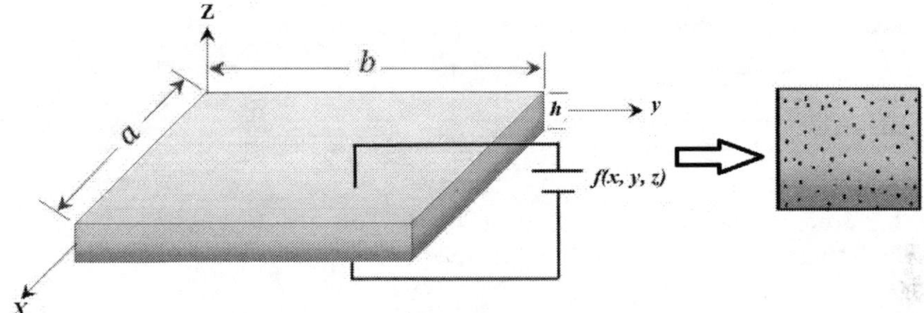

Fig. 1. Nano-plate FG Geometric of porosity.

The Young's modulus (E) and content density (ρ) equations of the FG nano-plate made from a mixture of two metal and ceramic materials can be represented by the distribution of the Power Rule (P-FGM) as:

$$E(z) = E_m \left(1 - V - \frac{\xi}{2} \right) + E_c \left(V - \frac{\xi}{2} \right) \quad \text{and} \quad V(z) = \left(\frac{1}{2} + \frac{z}{h} \right)^p \tag{6a}$$

$$\rho(z) = \rho_m \left(1 - V - \frac{\xi}{2} \right) + \rho_c \left(V - \frac{\xi}{2} \right) \quad \text{and} \quad V(z) = \left(\frac{1}{2} + \frac{z}{h} \right)^p \tag{6b}$$

Represent an imperfect FG with a density porosity proportion, ξ, ($0 \leq \xi \leq 1$) distributed equally between metal and ceramic in this sample. In the present analysis, the power law function is used. Distribution of power rule (P-FGM) in accordance with basic constituent combination rules.

That shear deformation plate hypothesis is appropriate for the displacements in the present analysis[27]:

$$u(x, y, z) = u_0(x, y) - z \frac{\partial w_b}{\partial x} + f(z) \frac{\partial w_s}{\partial x}$$

$$v(x, y, z) = v_0(x, y) - z \frac{\partial w_b}{\partial y} + f(z) \frac{\partial w_s}{\partial y}$$

$$w(x, y, z) = w_b(x, y) + w_s(x, y)$$

$$(7)$$

Where u_0 and v_0 represent the displacement functions of the plate's middle surfaces. The representative shape function which denotes the distribution of transverse shear stress or strain along the thickness of the plate is also $f(z)$. We have in this study

$$f(z) = \sin \left(\frac{\pi}{h} z \right) e^{1/2 \cos(\pi z / h)} + \frac{\pi}{2h} z \tag{8}$$

The field of strain is computed by:

$$
\begin{Bmatrix} \varepsilon_x \\ \varepsilon_y \\ \gamma_{xy} \end{Bmatrix} = \begin{Bmatrix} \dfrac{\partial u_0}{\partial x} \\[2mm] \dfrac{\partial v_0}{\partial y} \\[2mm] \dfrac{\partial u_0}{\partial y} + \dfrac{\partial v_0}{\partial x} \end{Bmatrix} - z \begin{Bmatrix} \dfrac{\partial^2 w_b}{\partial x^2} \\[2mm] \dfrac{\partial^2 w_b}{\partial y^2} \\[2mm] 2\dfrac{\partial^2 w_b}{\partial x \partial y} \end{Bmatrix} + f(z) \begin{Bmatrix} \dfrac{\partial^2 w_s}{\partial x^2} \\[2mm] \dfrac{\partial^2 w_s}{\partial y^2} \\[2mm] 2\dfrac{\partial^2 w_s}{\partial x \partial y} \end{Bmatrix}, \quad \begin{Bmatrix} \gamma_{yz} \\ \gamma_{xz} \end{Bmatrix} = \dfrac{df(z)}{dz} \begin{Bmatrix} \dfrac{\partial w_s}{\partial y} \\[2mm] \dfrac{\partial w_s}{\partial x} \end{Bmatrix}, \varepsilon_z = 0
$$

$$(9)$$

For a linear elastic plate and isotropic, stress-strain relationships

$$
\begin{Bmatrix} \sigma_x \\ \sigma_y \\ \tau_{xy} \end{Bmatrix} = \begin{bmatrix} Q_{11} & Q_{12} & 0 \\ Q_{12} & Q_{22} & 0 \\ 0 & 0 & Q_{66} \end{bmatrix} \begin{Bmatrix} \varepsilon_x \\ \varepsilon_y \\ \gamma_{xy} \end{Bmatrix} \quad \begin{Bmatrix} \tau_{yz} \\ \tau_{zx} \end{Bmatrix} = \begin{bmatrix} Q_{44} & 0 \\ 0 & Q_{55} \end{bmatrix} \begin{Bmatrix} \gamma_{yz} \\ \gamma_{zx} \end{Bmatrix}
$$

$$(10)$$

Where the stress and strain components are (σ_x, σ_y, τ_{xy}, τ_{yz}, τ_{yx}) and (ε_x, ε_y, γ_{xy}, γ_{yz}, γ_{yx}) respectively. Using the material properties described in Eq.(5), it is possible to express the stiffness coefficients, Q_{ij}, as

$$
Q_{11} = Q_{22} = \frac{E(z)}{1-v^2}, \quad Q_{12} = \frac{v\,E(z)}{1-v^2}, \quad Q_{44} = Q_{55} = Q_{66} = \frac{E(z)}{2(1+v)},
$$

$$(11)$$

To obtain the equations of motion, Hamilton's theory is used herein. The theory can be described as Reddy [28] in empirical terms.

$$
\int_0^T (\delta U - \delta K)\,dt = 0
$$

$$(12)$$

The variance of the plate's strain energy can be written as

$$
\delta U = \int_{-h/2}^{h/2} \int_A \left[\sigma_x \delta\,\varepsilon_x + \sigma_y \delta\,\varepsilon_y + \tau_{xy} \delta\,\gamma_{xy} + \tau_{yz} \delta\,\gamma_{yz} + \tau_{xz} \delta\,\gamma_{xz} \right] dA\,dz
$$

$$(13)$$

It is necessary to write the variance in kinetic energy of the plate as

$$
\delta K = \int_{-h/2}^{h/2} \int_A \rho(z)(\ddot{u}\,\delta u + \ddot{v}\,\delta v + \ddot{w}\,\delta w)\,dA\,dz
$$

$$(14)$$

Where the convention of the dot-superscript indicates the distinction with respect to the time variable t; and ($I_1, I_2, I_3, I_4, I_5, I_6$) are mass inertias specified as

$$
\left(I_1, I_2, I_3, I_4, I_5, I_6\right) = \int_{-h/2}^{h/2} \left(1, z, z^2, f(z), zf(z), f(z)^2\right) \rho(z)\,dz
$$

$$(15)$$

Using the integral by section and after simplification, the equilibrium equations associated with the present formulation for the nonlocal plate:

$$
\delta u : \quad \frac{\partial N_x}{\partial x} + \frac{\partial N_{xy}}{\partial y} = I_1 \ddot{u}_0 - I_2 \frac{\partial \ddot{w}_b}{\partial x} - I_4 \frac{\partial \ddot{w}_s}{\partial x}
$$

$$
\delta v : \quad \frac{\partial N_{xy}}{\partial x} + \frac{\partial N_y}{\partial y} = I_1 \ddot{v}_0 - I_2 \frac{\partial \ddot{w}_b}{\partial y} - I_4 \frac{\partial \ddot{w}_s}{\partial y}
$$

$$\delta w_b: \quad \frac{\partial^2 M_x^b}{\partial x^2} + 2\frac{\partial^2 M_{xy}^b}{\partial x \partial y} + \frac{\partial^2 M_y^b}{\partial y^2} = I_1(\ddot{w}_b + \ddot{w}_s) + I_2\left(\frac{\partial \ddot{u}}{\partial x} + \frac{\partial \ddot{v}}{\partial y}\right) - I_3\left(\frac{\partial^2 \ddot{w}_b}{\partial x^2} + \frac{\partial^2 \ddot{w}_b}{\partial y^2}\right) - I_5\left(\frac{\partial^2 \ddot{w}_s}{\partial x^2} + \frac{\partial^2 \ddot{w}_s}{\partial y^2}\right)$$

$$\delta w_s: \quad \frac{\partial^2 M_x^s}{\partial x^2} + 2\frac{\partial^2 M_{xy}^s}{\partial x \partial y} + \frac{\partial^2 M_y^s}{\partial y^2} + \frac{\partial S_{xz}^s}{\partial x} + \frac{\partial S_{yz}^s}{\partial y} = I_1(\ddot{w}_b + \ddot{w}_s) + I_4\left(\frac{\partial \ddot{u}}{\partial x} + \frac{\partial \ddot{v}}{\partial y}\right) - I_5\left(\frac{\partial^2 \ddot{w}_b}{\partial x^2} + \frac{\partial^2 \ddot{w}_b}{\partial y^2}\right) - I_6\left(\frac{\partial^2 \ddot{w}_s}{\partial x^2} + \frac{\partial^2 \ddot{w}_s}{\partial y^2}\right)$$

$$(16)$$

Analytical Solutions for Nonlocal Plates

We assume the following type of solution for (u, v, w_b, w_s) following the Navier solution procedure.

$$\begin{Bmatrix} u \\ v \\ w_b \\ w_s \end{Bmatrix} = \begin{Bmatrix} U_{mn}e^{i\omega t}\cos(\alpha x)\sin(\beta y) \\ V_{mn}e^{i\omega t}\sin(\alpha x)\cos(\beta y) \\ W_{bmn}e^{i\omega t}\sin(\alpha x)\sin(\beta y) \\ W_{smn}e^{i\omega t}\sin(\alpha x)\sin(\beta y) \end{Bmatrix}$$

$$(17)$$

Where the parameters U_{mn}, V_{mn}, W_{bmn}, and W_{smn} are arbitrary. The eigen frequency correlated with (m, n)th eigenmode is ω, and mode numbers are $\alpha = m\pi/a$ and $\beta = n\pi/b$, «m» and «n». A system of equations may be combined as:

$$\left([K] - \lambda\omega^2[M]\right)\{\Delta\} = \{0\}$$

$$(18)$$

Where, respectively, [K] and [M], stiffness and mass matrices, and defined as:

$$\left(\begin{vmatrix} a_{11} & a_{12} & a_{13} & a_{14} \\ a_{12} & a_{22} & a_{23} & a_{24} \\ a_{13} & a_{23} & a_{33} & a_{34} \\ a_{14} & a_{24} & a_{34} & a_{44} \end{vmatrix} - \lambda\omega^2 \begin{vmatrix} m_{11} & 0 & 0 & 0 \\ 0 & m_{22} & 0 & 0 \\ 0 & 0 & m_{33} & m_{34} \\ 0 & 0 & m_{34} & m_{44} \end{vmatrix}\right)\begin{Bmatrix} U_{mn} \\ V_{mn} \\ W_{bmn} \\ W_{smn} \end{Bmatrix} = \begin{Bmatrix} 0 \\ 0 \\ 0 \\ 0 \end{Bmatrix}$$

$$(19)$$

in which

$$a_{11} = -\left(A_{11}\lambda^2 + A_{66}\mu^2\right)$$

$$a_{12} = -\lambda\mu\left(A_{12} + A_{66}\right)$$

$$a_{13} = \lambda\,[\,B_{11}\lambda^2 + (B_{12} + 2B_{66})\,\mu^2\,]$$

$$a_{14} = \lambda\,[\,B_{11}^s\lambda^2 + (B_{12}^s + 2B_{66}^s)\,\mu^2\,]$$

$$a_{22} = -\left(A_{66}\lambda^2 + A_{22}\mu^2\right)$$

$$a_{23} = \mu\,[(B_{12} + 2B_{66})\,\lambda^2 + B_{22}\mu^2\,]$$

$$a_{24} = \mu\,[(B_{12}^s + 2B_{66}^s)\,\lambda^2 + B_{22}^s\mu^2\,]$$

$$a_{33} = -\left(D_{11}\lambda^4 + 2(D_{12} + 2D_{66})\lambda^2\mu^2 + D_{22}\mu^4\right)$$

$$a_{34} = -\left(D_{11}^s\lambda^4 + 2(D_{12}^s + 2D_{66}^s)\lambda^2\mu^2 + D_{22}^s\mu^4\right)$$

$$a_{44} = -\left(H_{11}^s\lambda^4 + 2(H_{12}^s + 2H_{66}^s)\lambda^2\mu^2 + H_{22}^s\mu^4 + A_{55}^s\lambda^2 + A_{44}^s\mu^2\right)$$

$$(20)$$

$$m_{11} = -I_1; \quad m_{12} = 0; \quad m_{13} = \lambda I_2; \quad m_{14} = \lambda I_4$$

$$m_{21} = 0; \quad m_{22} = -I_1; \quad m_{23} = \mu I_2; \quad m_{24} = \mu I_4$$

$$m_{31} = \lambda I_2; \quad m_{32} = \mu I_2; \quad m_{33} = -\left[I_1 + I_3\,(\lambda^2 + \mu^2)\right]; \quad m_{34} = -I_5\,(\lambda^2 + \mu^2)$$

$$m_{41} = \lambda I_4; \quad m_{42} = \mu I_4; \quad m_{43} = -I_5\,(\lambda^2 + \mu^2); \quad m_{44} = -I_6\,(\lambda^2 + \mu^2)$$

$$\lambda = 1 + \eta\left(\alpha^2 + \beta^2\right)$$

$$(21)$$

Numerical Results and Discussions

The values of material properties for FGM nano-plates used in the present study are set out in this part.

Metal (SUS304): E_m=201.04GPa, ρ=8166kg/m^3, υ=0.3.

Ceramics (Si$_3$N$_4$) : E_c=348.43GPa, ρ=2370kg/m^3, υ =0.3.

For convenience, the non-dimensional frequency parameter used in this study is:

$$\hat{\omega} = \omega h \sqrt{\frac{\rho}{G}}$$

(22)

In the present analysis, dimensionless natural frequencies values are used for simple supported FG nano-plate conditions for different nonlocal parameter values and the plate aspect ratio as compared in Table 1.

The numerical findings from the new analysis are in very close alignment with the current solutions in table 1. It is noted that the nonlocal theory has values of natural frequency lower than the theory of local elasticity. A comparison of the findings of the dimensionless frequency observed in the present theory and with the results (Zargaripoor et al[26]) is given in Table 2 for the different vibration modes of the clearly assisted nano-plate and the various non-local parameters " η ". This table illustrates a strong consensus between the findings, as can be shown.

Table 1. Comparisons of dimensionless frequency ($\hat{\omega}$) of square and rectangular FG nanoplate (a/h=10, P=0)

a/b	Method	Nonlocal Parameter (η)		
		0	1	2
	Natarajan [23]	0,0929	0,0849	---
1	Aghababaei [25]	0,0930	0,0850	0,0788
	Zargaripoor [26]	0,0930	0,0850	0,0788
	Present	0,0931	0,0851	0,0788
	Natarajan [23]	0,0590	0,0556	---
2	Aghababaei [25]	0,0589	0,0556	0,0527
	Zargaripoor [26]	0,0589	0,0556	0,0527
	Present	0,0589	0,0556	0,0527

Table 2. Comparison of natural frequency of square FG nano-plate (a/h=10, a=b).

Power law index "p"	Nonlocal Parameter " η "	Mode 1 (1,1)		Mode 2 (1,2)	
		Zargaripoor et al [26]	Present	Zargaripoor et al [26]	Present
	0	0,0930	0,0931	0,2225	0,2225
0	1	0,0850	0,0851	0,1820	0,1820
	2	0,0788	0,0789	0,1578	0,1578
	0	0,0552	0,0548	0,1310	0,1308
1	1	0,0504	0,0501	0,1072	0,1070
	2	0,0467	0,0464	0,0930	0,0928
	0	0,0444	0,0442	0,1052	0,1052
5	1	0,0405	0,0404	0,0861	0,0861
	2	0,0376	0,0374	0,0747	0,0746

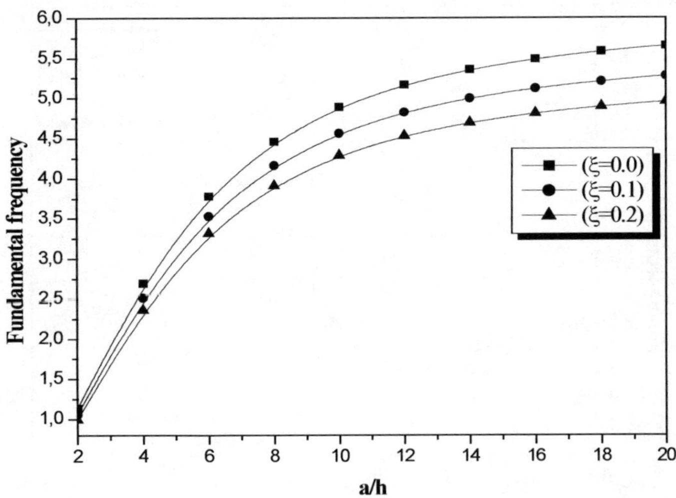

Fig. 2. Variation of the fundamental frequency as a function of the thickness ratio and the porosity coefficient.

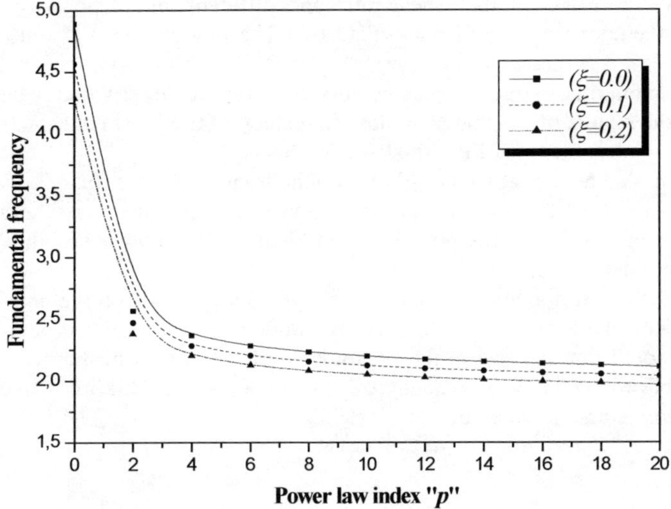

Fig. 3. Fundamental frequency in function of the Power-law index with (a/h = 10) and porosity coefficient (ξ).

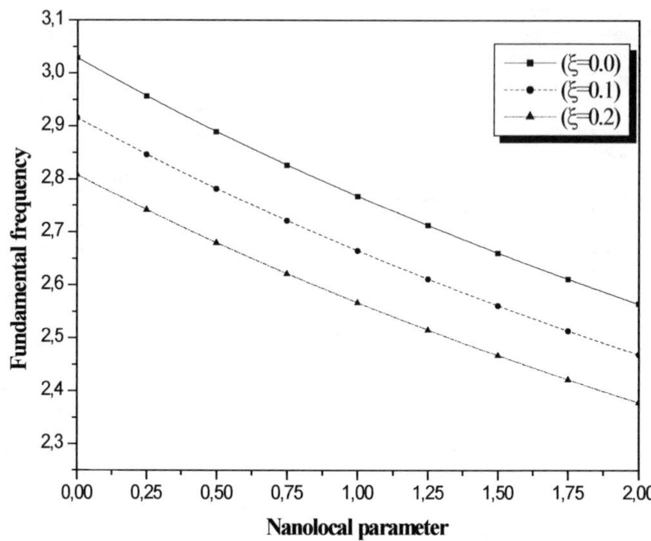

Fig. 4. Variation of fundamental frequency in function of a non-local parameter.

Figure 2, shows the effect of the aspect ratio and different values porosity coefficient (ξ) on dimensionless frequency responses of nano-plate FG. The nonlocal ($\eta = 1$ nm) results are given. The power-law index is assumed to be constant, $p = 0$. The increase in coefficient of increasing porosity on the fundamental frequency became very clear for the larger values of the ratio (a/h). The curves show that the results of the dimensionless frequency of the local nano-plates are greater than the non-local nano-plates because of the small-scale effects.

The variance of the fundamental non-dimensional frequency as a function of the power-law index and the porosity coefficient values (ξ) for the non-local nano-plate ($\eta = 2$ nm) are presented in Figure 3. It can be shown that as the porosity coefficient and the power law index rise, the simple frequency parameter decreases.

The variance of the non-dimensional fundamental frequency based on the non-local parameter is seen in Figure 4 with the aspect ratio a/h = 10, the power-law index $p = 0$, and the coefficient of porosity (ξ). It is noted that the rise in the coefficients of porosity contributes to a decrease in the frequency of simple dimensionless frequencies. For higher values, the influence of the non-local parameter (η) is simple and noticeable.

Conclusion

In this paper the study of free vibration of porous nano-plates FG using non-local principle and high order shear and normal deformation theory. The inclusion of a nano-plate permitted all the effects of different parameters such as thickness-length ratio, aspect ratio, power-law exponent and porosity factor on the natural frequencies of FG nano-plates to be studied and analyzed. The findings obtained from the dimensionless frequency analysis are in rather close alignment with the results reported by Natarajan [23], Aghababaei [25] and the results presented by Zargaripoor et al[26] of the finite element system approach and those expected to appear in the literature. It can be deduced from this study that:

- If the aspect ratio of the simply supported FG nano-plate increases, the non-dimensional frequency increases.
- The frequency ratio decreases with the increase of the non-local parameter for a different mode of vibration.
- The dimensionless frequency decreases with increasing power law exponent.

References

[1] Miyamoto, M., Kaysser, W.A., Rabin, B.H., Functionally Graded Materials Design, Processing and Applications, (1999).

[2] Suresh, S., Mortensen, A., Fundamentals of Functionally Graded Materials, IOM Communications Ltd., London, (1998).

[3] Öchsner, A., Murch, G.E. and Lemos, M.J.S., Cellular and Porous Materials, WILEY-VCH, 398-417, (2008).

[4] Hadj Mostefa. A., Merdaci. S, and Mahmoudi. N., An Overview of Functionally Graded Materials «FGM», Proceedings of the Third International Symposium on Materials and Sustainable Development, ISBN 978-3-319-89706-6, 267–278, (2018).

[5] Zhu, J. Lai, Z. Yin, Z. Jeon, J. and Lee, S., Fabrication of ZrO_2–NiCr functionally graded material by powder metallurgy, Mater. Chem. Phys, 68(1-3), 130-135, (2001).

[6] Wattanasakulpong, N., Prusty, B.G., Kelly, D.W. and Hoffman, M., Free vibration analysis of layered functionally graded beams with experimental validation, Mater. Des, 36, 182-190, (2012).

[7] Shimpi, R., Patel, H., Free vibrations of plate using two variable refined plate theory, J. Sound Vib, 296, 979–999, (2006).

[8] Jha, D.K., Kant, T. and Singh, R.K., Higher order shear and normal deformation theory for natural frequency of functionally graded rectangular plates, Nucl. Eng. Des., 250, 8–13, (2012).

[9] Wattanasakulpong, N. and Ungbhakorn, V., Linear and nonlinear vibration analysis of elastically restrained ends FGM beams with porosities, Aerosp.Sci. Technol, 32(1), 111-120, (2014).

[10] Merdaci, S., Belmahi, S., Belghoul, H., Hadj Mostefa, A., Free Vibration Analysis of Functionally Graded Plates FG with Porosities, International Journal of Engineering Research & Technology, 8(03), 143-147, (2019).

[11] Merdaci, S., Free Vibration Analysis of Composite Material Plates "Case of a Typical Functionally Graded FG Plates Ceramic/Metal" with Porosities, Nano Hybrids and Composites (NHC), 25, 69-83, (2019).

[12] Merdaci. S, Hadj Mostefa .A, Merazi .M, Belghoul .H, Hellal .H, Boutaleb .S, "Effects of even pores distribution of functionally graded plate porous rectangular and square", Procedia Structural Integrity, 26, 35–45, (2020).

[13] Merdaci. S, Hadj Mostefa. A, Beldjelili. Y, Merazi.M, Boutaleb. S, Hellal. H, "Analytical solution for static bending analysis of functionally graded plates with porosities", Frattura ed Integrità Strutturale, 55, 65-75, (2021).

[14] Merdaci S; Hadj Mostefa A; Merazi M; Belghoul H; Boutaleb S; Hellal H, "Free Vibration Analysis of Ceramic-Metal Functionally Graded rectangular Solar Plates with Porosities Using of High Order Shear Theory: Solar Plate FG Composed of (Al/Al_2O_3) and (Al/ZrO_2) Influence by Porosity", Institute of Electrical and Electronics Engineers (IEEE - Journals & Conference Proceedings), 1-5, (2021).

[15] Daneshmehr, A., Rajabpoor, A., Hadi, A., Size dependent free vibration analysis of nanoplates made of functionally graded materials based on nonlocal elasticity theory with high order theories, International Journal of Engineering Science, 95, 23-35 (2015).

[16] Karami, B., Shahsavari, D., Janghorban, M., Wave propagation analysis in functionally graded (FG) nanoplates under in-plane magnetic field based on non-local strain gradient theory and four variable refined plate theory, Mech. Adv. Mat. Struct, (2017).

[17] Shahsavari,D., Karami, B., Mansouri, S., Shear buckling of single layer graphene sheets in hygrothermal environment resting on elastic foundation based on different nonlocal strain gradient theories, Eur. J. Mech. A, Solids, (2017).

[18] Eringen, A.C., Edelen, D.G.B.,On nonlocal elasticity, Int. J.Eng.Sci, 10, 233–248, (1972).

[19] Eringen, A.C., On differential equations of nonlocal elasticity and solutions of screw dislocation and surface waves, J. Appl.Phys, 54(9), 4703–4710, (1983).

[20] Lu, P., Zhang, P., Lee, H., Wang, C., Reddy, J., Actes de la Royal Society A, 463, 3225–3240, (2007).

[21] Murmu, T., Pradhan, S., Physica E: Systèmes à basse dimension et Nanostructures ,41(8), 1628–1633, (2009).

[22] Merdaci. S, Hadj Mostefa. A, Boutaleb. S, Hellal. H, "Free Vibration Analysis of Functionally Graded FG Nano-plates with Porosities", Journal of Nano Research, Vol.64, pp 61-74, (2020).

[23] Natarajan, S., Chakraborty, S., Thangavel, M., Bordas, S., Rabczuk, T., Size-dependent free flexural vibration behavior of functionally graded nanoplates, Computational Materials Science, 65, 74-80, (2012).

[24] Salehipour, H., Nahvi, H., Shahidi, A., Exact closed-form free vibration analysis for functionally graded micro/nano plates based on modified couple stress and three-dimensional elasticity theories, Composite Structures, 124, 283-291, (2015).

[25] Aghababaei, R., Reddy, J.N., Nonlocal third-order shear deformation plate theory with application to bending and vibration of plates, Journal of Sound and Vibration, 326, 277–289, (2009).

[26] Zargaripoor, A., Daneshmehr, A., Isaac Hosseini, I., Rajabpoor, A., Free vibration analysis of nanoplates made of functionally graded materials based on nonlocal elasticity theory using finite element method, Journal of Computational Applied Mechanics, 49(1), 86-101, (2018).

[27] Merdaci, S., Tounsi, A., Houari, M.S.A., Mechab, I., Hebali, H., Benyoucef, S., Two new refined shear displacement models for functionally graded sandwich plates, Arch Appl Mech, 81, 1507-1522, (2011).

[28] Reddy, J. N., and Phan, N. D. "Stability and vibration of isotropic, orthotropic and laminated plates according to a higher-order shear deformation theory, J. Sound Vibrat, 98, 157–170, (1985).

Nano Hybrids and Composites
ISSN: 2297-3370, Vol. 33, pp 93-103
© 2021 Trans Tech Publications Ltd, Switzerland

Submitted: 2021-06-29
Revised: 2021-08-21
Accepted: 2021-09-01
Online: 2021-10-11

Far Infrared Laser Detector Based on Multi-Walled Carbon Nanotubes and Blend of (Polyaniline - Polymethyl Methacrylate) Polymers with Methyl Blue Dye for Photoconductive Applications

Wasan R. Saleh[1,a*], Salma M. Hassan[1,b], Samar Y. Al-Dabagh[2,c],
Marwa A. Al-Azzawi[1,d]

[1]Department of Physics, College of Science, University of Baghdad, Baghdad, Iraq

[2]Department of Physics, College of Science for Women, University of Baghdad, Baghdad, Iraq

[a*]wasan_alazawi@yahoo.com, [b]salma.mhammed@yahoo.com, [c]samarjam2002@yahoo.com,
[d]marwaabdulrahman90@gmail.com

Keywords: Carbon nanotubes, Infrared detector, Polyaniline polymer, Polymethyl methacrylate polymer, Methyl Blue dye.

1. Abstract: Infrared photoconductive detectors working in the far-infrared region and room temperature were fabricated. The detectors were fabricated using three types of carbon nanotubes (CNTs); Multi-Walled Carbon Nanotubes (MWCNTs), carboxyl Multi-Walled Carbon Nanotubes (COOH-MWCNTs), and short-MWCNTs. The carbon nontubes suspension is deposited by dip coating and drop–casting techniques to prepare thin films of CNTs. These films were deposited on porous silicon (PSi) substrates of n-type silicon (Si). The current-voltage (I-V) characteristics and the figures of merit of the fabricated detectors were measured at a forward bias voltage of 3 and 5 volts as well as at dark and under illumination by infrared (IR) radiation from a carbon dioxide (CO_2) laser of 10.6 μm wavelengths and power of 2.2 Watt. The responsivity and figures of merit of the photoconductive detector are improved by coating the MWCNTs films with a thin layer of a blend (polyaniline - polymethyl methacrylate) polymer with methylene blue dye. The coated MWCNTs films showed better performances, so this type of coating can be considered as a surface treatment of the detector film, which highly increased the responsivity and specific detectivity of the fabricated IR laser detector-based MWCNTs. The photocurrent response for the coated films was increased about 25 times than that for uncoated films. The results proved the role of the polymer in the enhancement of the performance of the IR photoconductive detectors.

2. Introduction:

Infrared (IR) detection is of tremendous interest in industrial, scientific and military technologies and also in daily life. Infrared detectors are used in environmental monitoring, remote controls, optical communication, health care (thermography), astronomy, night surveillance, guided missile technology and many other latest technologies like self-driving cars [1]. Developing IR detectors with high performance is very important. It is possible to divide IR detectors into two categories: photon and thermal detectors [2]. Photoconductive (PC) detectors depends on the rise in the electrical conductivity due to an increase in the number of free carriers produced by the absorption of photons (current generation). Photoconductive effect results from direct conversion of incident photons into conducting electrons within a material. Detectors based on these effects are called photon detectors, because they directly convert photons into conducting electrons; no intermediate process is involved [3]. PC detectors usually own high-frequency responses; however, they also have a high signal-to-noise ratio [4].
New materials have been designed to improve the efficiency of IR detectors [5], but low noise and high detectivity are still difficult to produce. Due to their special band composition, outstanding electronic and optoelectronic properties and super-mechanical and chemical stability, carbon nanotubes (CNTs) are promising candidates for possible IR detectors [6]. Nowadays, CNTs by taking advantage of its special structure, it can be used to create IR detectors that have the ability to outperform conventional detectors. The photo-behavior of carbon nanotubes (CNTs) has attracted

great attention due to possible applications of their unique hollow cylindrical shape, electronic and optoelectronic properties, suitable band gap, mechanical and chemical stability [7].

Many materials and techniques can be used to enhance the properties of photoconductive detectors. Porous materials are considered of the most important construction. Understanding the pore structure of these materials is important because they provide insight into both the microstructure and material efficiency [8]. Porous silicon (PSi) consists of a network of nanoscale wires and voids that are created by photo-chemically etching crystalline silicon wafers. Their structure consists of few nanometers of silicon particles divided by voids in dimension. They are used in many chemical and biological sensing applications due to their large specific surface area (200-800 m^2/cm^3) [9].

Recently, some studies have focused on using polymer composites and blending technology to obtain hybrid materials with specific functionality. One such class of material is conducting polymers blend, which is a blend of an insulating polymer with a conducting polymer, and it has shown enhanced optical, electrical, and dielectric properties [10, 11]. PANI is a very important conducting polymer, which is being investigated in recent years due to its low cost, nontoxicity, reversibility, good environmental stability, and high intrinsic redox properties. Generally, the carbon nanotube/PANI composite was studied for its unique optical, electrical, and electrochemical properties [12].

New materials have been developed to improve the performance of IR detectors, but it is still challenging to fabricate a low noise, high detectivity detector [5]. CNTs are favorable materials for IR detectors because of their unique properties [6]. Nanocomposites of CNTs and conductive polymer (CP) have created high research interest, to improve their electrical conductivity and mechanical properties [13, 14]. The polymer composites, metal oxide decoration, metal, and different functional groups were introduced in CNTs to improve sensitivity and detectivity and for various applications [15].

In the last years, the researches [16-22] focused on the response of carbon nanotubes in the infrared region. These researches indicated that carbon nanotubes CNTs have a potential application in IR detection and exhibit strong infrared light absorption with broad band and fast light responses.

In this work, infrared photoconductive detectors working in the far-infrared region and room temperature were fabricated. Then coated the CNTs films with a thin layer of a blend (polyaniline - polymethyl methacrylate) polymer and methylene blue dye, in order to improve the detector characteristics.

3. Materials and Methods:

Raw materials:

Three types of MWCNTs were used in this work; multi-walled carbon nanotubes (MWCNTs) of length 10-30 μm and average diameter 5-10 nm, functionalized multi-walled carbon nanotubes (COOH-MWCNTs) of length 50μm and average diameter 3-5 nm with 2.56 wt% COOH content (both supplied by Neutrino Company), and short thin MWCNTs functionalized with 0.05% NH_2 (supplied by NANOCYL S.A.) of purity > 95.0%, the average diameter of the tubes was 9.5 nm, and the average length <1.0 μm.

The solutions in this work were: N-N-Dimethylformamide (DMF) solution (from OminSolv Company) which was used to prepare CNTs suspension, Polymethyl methacrylate polymer (PMMA) with the formula $(C_5O_2H_8)N$ (supplied by Sejong Enterprise Company,) Polyanilin polymer (PANI) with the formula $-[(B\text{-}NH\text{-}B\text{-}NH)_Y (B\text{-} N=Q=N)_{1-Y}]_{N^-}$ which was homemade at room temperature, and Methylene Blue dye (MB) with the formula $C_{16}H_{18}N_3SCl$ (supplied by Science Company).

Method:

The procedure of fabrication of the IR photoconductive detector was described in detail in [16, 17]. The steps are as followed: to prepare the porous silicon (PSi) structure, the photochemical etching process was used. For this purpose, the n-type crystalline silicon wafer of 0.05 Ω. cm resistivity was used as a substrate and soaked in hydrofluoric acid (HF) with a concentration of 10% in a Teflon beaker. Photons from a 250Watt tungsten lamp were focused on Si wafer using a convex lens of 5 cm focal length. The irradiation time necessary to create the PSi structure was 12 minutes.

Each type of carbon nanotubes (CNTs) was dispersed separately in 25 ml of DMF and stirred for 15 min which is important for the reaction and breaking up the CNTs agglomerates so as to have a homogenous suspension. CNTs suspension was deposited on the PSi substrate at room temperature by two techniques: drop-casting and dip coating techniques. The dip-coating withdrawal speed was 1mm/min.

The detector is composed of a CNTs film on which two aluminum electrodes (0.9 cm apart) were deposited using an evaporation technique with the aid of a micro mask of 0.4 mm electrode width, as illustrated in Fig. (1a). The final arrangement of the fabricated IR photoconductive detector is illustrated in Fig. (1b and c).

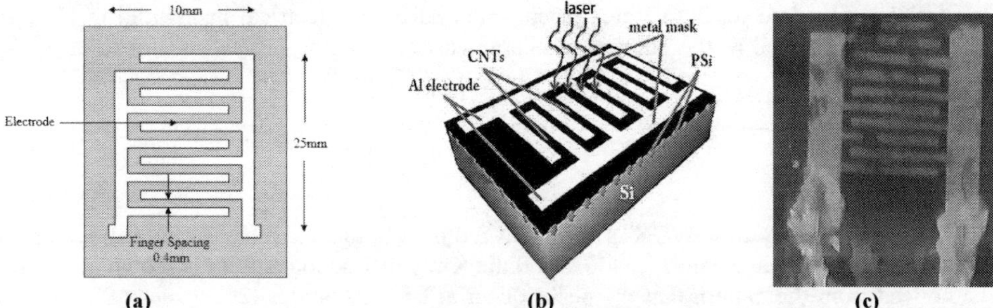

Fig. 1: (a) Illustration of the used electrode (IDE) masks and (b, c) schematic diagram and photo-plate for the final arrangement of the fabricated PC detector [16, 17].

The performance of the fabricated IR PC detectors was measured firstly for the detectors with CNTs film only. The second step was to coat the CNTs films with PANI, PMMA and MB dye, in order to improve the performance of the fabricated detector. For this purpose, (0.1, 1, and 0.0005) g of PANI, PMMA, and MB dye respectively were separately dissolved in 10 ml of DMF and stirred for (3 h, 1h, and 5min) respectively. Then a blend of (PANI-PMMA) polymers is mixed together with MB dye and stirred before deposited on the CNTs film. The concentration of MB dye was 1.56x10-4 M. The thickness of (PANI-PMMA) polymers blend and MB dye coating layer was 100 μm, measured using Michelson interferometer. The fabricated IR detector was connected in an electrical circuit (Fig. (2)), in order to study its I-V characteristic curves. This was done in the dark and under illumination using IR radiation from CO_2 laser of 10.6 μm wavelength and power of 2.2 Watt (W).

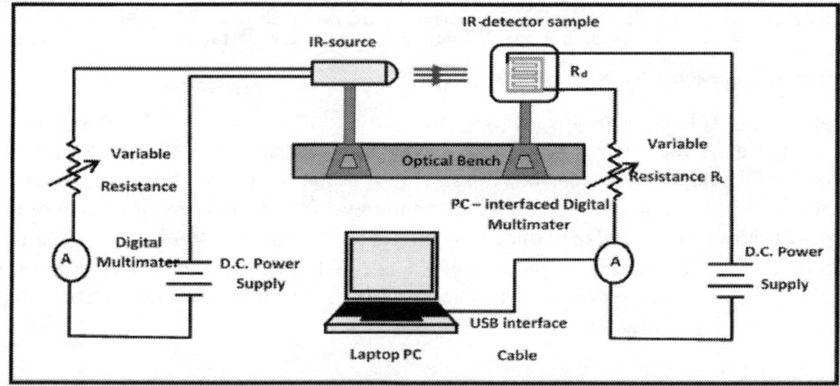

Fig. 2: Schematic diagram of the experimental setup, where R_d is the detector resistance and R_L is the load resistance (1 KΩ.).

4. Calculations:

The figures of merit such as responsivity R_λ, photocurrent gain G, noise equivalent power NEP, specific detectivity D* for the IR photoconductive detector were calculated depending on the value of the photo-current flowing between the electrodes I_{ph} and the dark current I_d by the following equation [3]:

$$R_\lambda = I_{ph} / P_{in} \tag{1}$$
$$G = (I_{ph}/e) \, (h\upsilon/P_{in}) \tag{2}$$
$$NEP = I_n / R_\lambda \tag{3}$$
$$D^* = R_\lambda \, (A \, \Delta \, f)^{1/2} \, / \, I_n \tag{4}$$
$$I_n = (\, 2e \, I_d \, \Delta F \,)^{1/2} \tag{5}$$

where P_{in} is the incident radiation power, h is the Planck's constant, υ is the frequency in Hz, I_n is the noise current, A is the detector active area in cm^2, and Δf is the electrical bandwidth in Hz. The photocurrent gain is defined as the number of charge carriers flowing per second for each photon absorbed per second.

5. Results and Discussion:

X-Ray Diffraction (XRD)

The crystalline structure of MWCNTs was analyzed by a Shimadzu 6000 X-ray diffractometer, using Cu Kα radiation of wavelength 1.5406A°and the X-ray diffraction pattern is shown in Fig. (3). It can be noticed from the pattern that the peaks occur at Bragg's angles (2θ) values of 26.02° and 42.79°. These angles are associated with diffraction peaks (002) and (100) which indicate that they have hexagonal (Wurtzite) polycrystalline structure, these result is in agreement with [23].

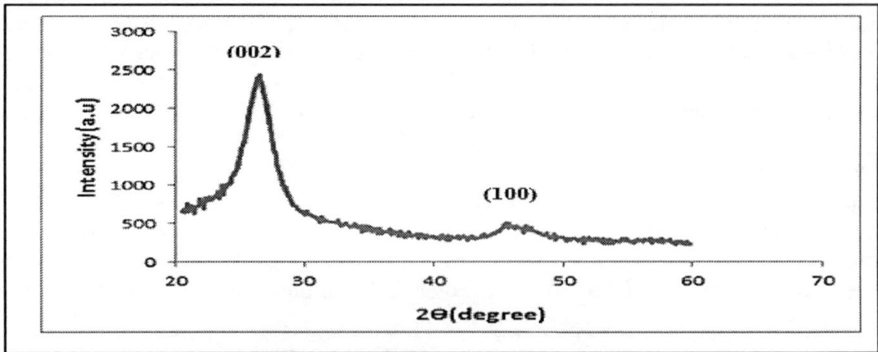

Fig. 3: X-ray diffraction pattern of the CNTs.

Atomic Force Microscope (AFM) of MWCNTs

The surface morphology of the carbon nanotube film was studied by using Atomic Force Microscope type AA3000 (AFM) supply by Angstrom Company. The morphology study helps in determination of the particles dimensions range of the carbon nanotube grain size and their statistical distribution. An AFM images of the surface morphology of the CNTs/ PSi is shown in Fig. (4 a, b), while Fig. (4c) shows the granularity distribution. It is clear from this figure that the film had a good and uniform surface homogeneity which gives a good indicate for deposition of the CNTs nanoparticles over the nanospikes of Si. The granular film shows higher surface area which is conductive for film-light interaction.

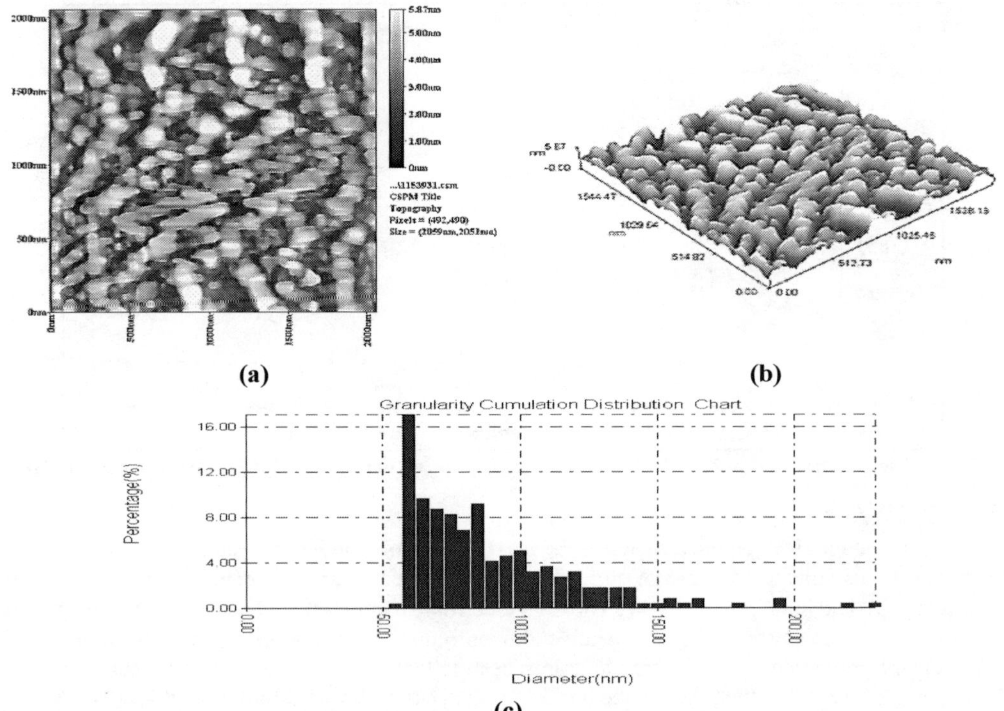

Fig. 4: (a) 2D, (b) 3D AFM images, and (c) particles size distribution for PSi/CNT film.

The average particle size determined from AFM is about 87.05 nm. The root mean square roughness (Sq) average roughness (Sa), and ten point height (Sz) for PSi/CNTs films were 1.24 nm, 1.04 nm and 5.86 nm respectively.

Fourier-Transform Infrared (FTIR) Spectrum

FT-IR Shimadzu spectrophotometer model 8300 was carried out to study the FTIR spectra for CNTs before and after coating with blend of (PANI-PMMA) polymers and MB dye in the range (400-4000) cm^{-1} are shown in Fig.s (5) and (6). The absorbance is plotted as a function of wave number. The figures show that the CNTs have good absorbance around 1000 cm^{-1} (10 µm). This is mean that these materials are very good candidate for IR applications, such as IR photoconductive detector.

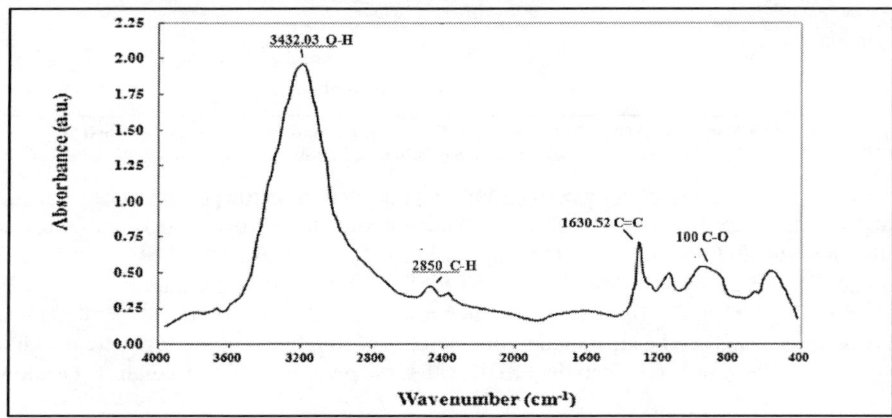

Fig. 5: FTIR absorbance spectrum of MWCNTs film.

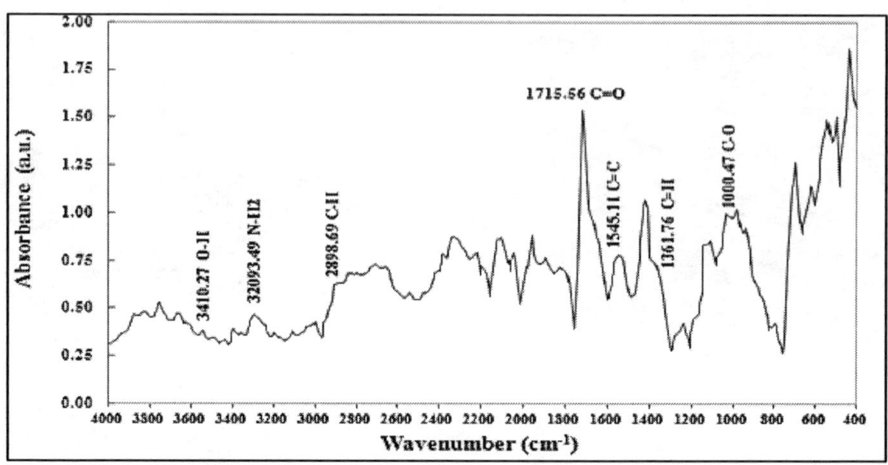

Fig. 6: Absorbance spectrum MWCNTs after coating with blend of (PANI-PMMA) polymers and MB dye.

I-V Characteristics

The current-voltage (I-V) characteristics of the fabricated photoconductive detectors were carried out at a forward bias voltage of (0.5-5) Volt in the dark and under illumination using CO_2 laser of 10.6 μm wavelength and input power of 2W. The used electronic bandwidth and detector active area were 1 Hz and $1cm^2$, respectively. Fig. (7) shows the current-voltage curves for the IR detector for MWCNTs-PSi prepared by dip coating and drop casting techniques using n-type PSi substrate.

The experiment revealed that, the device has low sensitivity to infrared radiation at low bias voltage (V< 2 Volt), which increases exponentially as the bias voltage (V > 2 Volt) increases. This is because the external shell of MWCNTs with low band gap could be broken and the current passing the CNTs dropped step by step by applying a certain voltage, according to Collin et al. [24].

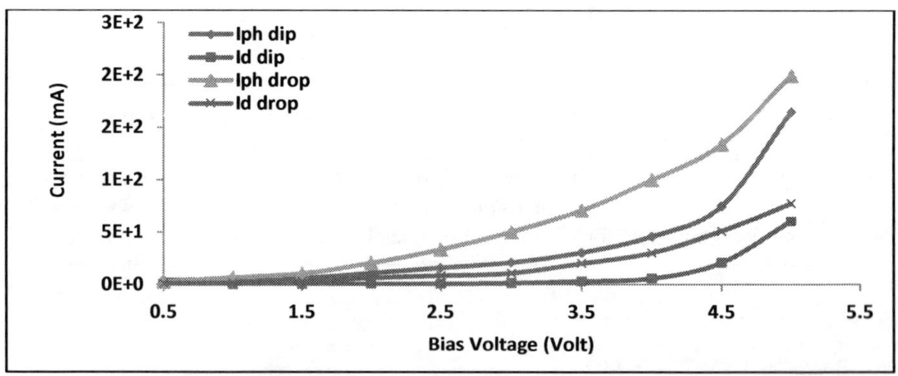

Fig. 7: The I-V characteristics of the MWCNTs-PSi IR detector using n-type Si substrate, the films deposited by dip coating and drop casting.

To understand the effect of etching on the sensitivity as well as on the performance of the detector, a comparison was made between two detectors, one without etching of the Si wafer (substrate), the other with the Si substrate etched with diluted HF solution (this produces the PSi). It is obvious from Fig. (8) that the photo-current of the detector with PSi substrate is larger than that for the detector with the not-etched substrate (Si wafer). The increase of I_{ph} after etching the substrate is about one order of magnitude. This can be attributed to the increase of roughness of the substrate which leads to an increase in the adhesion of CNTs to the PSi. Besides, the prepared PSi has a high surface to volume ratio.

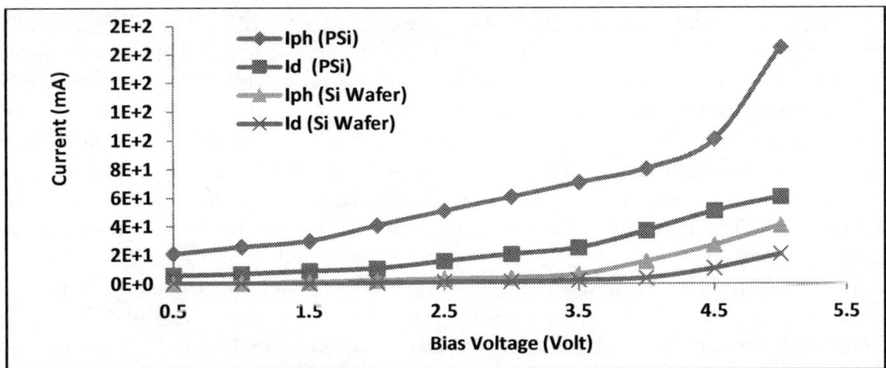

Fig. 8: The I-V characteristics of the MWCNTs-PSi IR detector using n-type Si substrate, the films deposited by dip coating technique on the etched substrate (PSi) and without etched substrate (Si wafer).

Fig. (9) shows the I-V curves for the detectors that were fabricated by depositing the MWCNTs, COOH-MWCNTs, and short MWCNTs films on the n-type PSi substrate by the drop-casting tecnique. The figure reveals that the sensitivity of the pure MWCNTs film is the best.

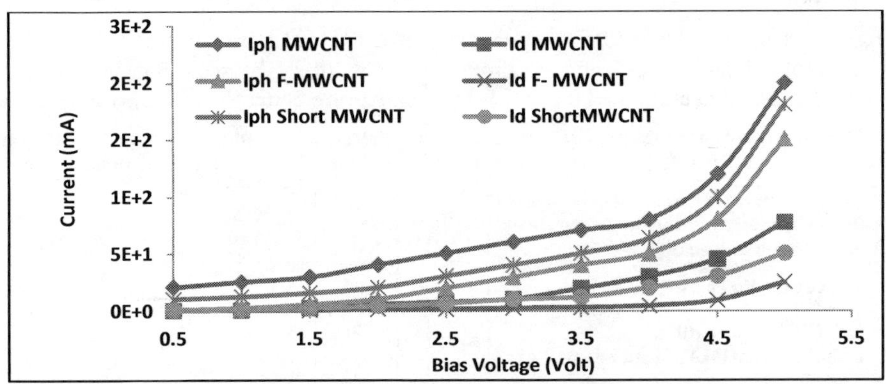

Fig. 9: The I-V characteristics of the MWCNTs, COOH-MWCNTs, and short MWCNTs PSi IR detectors using n-type Si substrate, the films deposited by drop casting technique.

To enhance the sensitivity and the figures of merit, the fabricated detector with the types of CNTs discussed in Fig. (9), were modefied by coating the CNTs film with a thin layer of (PANI-PMMA) blend and MB dye. The I-V curves of these films are shown in Fig. (10).

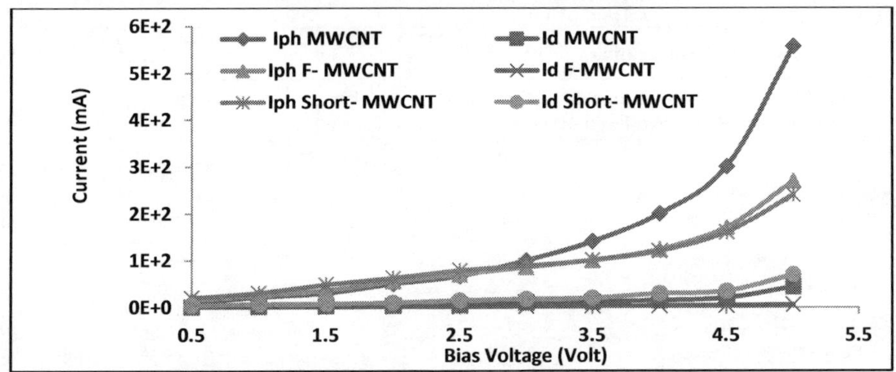

Fig. 10: The I-V characteristics for the MWCNTs, COOH-MWCNTs, and short-MWCNT PSi IR detectors prepared using n-type Si substrate, the films deposited by drop casting technique. After coating the CNTs layer with blend of (PANI- PMMA) polymers and MB dye.

From this figure, it can be noticed that coating the CNTs film with a layer of blend of (PANI-PMMA) polymers and MB dye caused a large increase in photo-current. This may be attributed to the IR absorption capability of the blend of (PANI-PMMA) polymers and MB dye that caused excited states in the polymer. The excited electrons leave the ground states (unoccupied orbital) at an energy level within the bandgap of the CNTs semiconductor. The excitation probability of an electron to the conduction band increases in these states, which may be considered as transition states for electrons of the valence band for CNTs, to transit to the conduction band. The reduction in the reflected parts of the incident IR radiation on the CNTs detectors is due to the matching index between the CNTs and the coated polymer layer. MWCNTs refractive index was found to be around 2.5 , so the reflection parts R_o of the incident infrared radiation is more than (18%) as determined using Fresnel's relation for normal incident $R_o = (n-1/n+1)^2$, where n is the refractive index. The refractive index of the blend coating layer was measured with Abbe refractometer to be equal to 1.433. This coating acts as a matching index layer with a refractive index equal to $\sqrt{n_{cnts}.n_{air}} \approx 1.58$, which is not far from the n value of the (PANI-PMMA) blend and MB dye. This matching index greatly decreases the reflectivity of infrared radiation on the surface of the carbon nanotubes film, which greatly increases the absorption of the incident radiation [25].

Figures of Merit

The figures of merit calculated from Eqs. (1-5) at a forward bias voltage of 3V and 5 V are tabulated in Tables 1 and 2, respectively. From the tables, it can be seen that, in most cases, the fabricated far IR laser detectors working at 5 V bias voltage are better than that working at 3V.

Table 1: Some figures of Merit for the CNTs IR fabricated detectors prepared using n-type PSi substrate and worked at 3V forward bias voltage, $\Delta f=1Hz$, and active area $A=1cm^2$. The laser source of 10.6 µm wavelength.

Sample Conditions: (CNTs type, deposition technique, with or without blend coating layer)	R_λ [A/W]	G	NEP [Watt]	D* [(cm.Hz$^{1/2}$)/W]
MWCNTs, dip coating	0.0006	2.005	9.05 x10^{-8}	1.10 x 10^{+7}
MWCNTs, dip coating, with (PANI-PMMA) blend and MB dye	0.0034	50	8.41 x10^{-13}	1.19 x 10^{+12}
MWCNTs, drop coating	0.0015	4.690	3.66 x 10^{-8}	2.73 x 10^{+7}
MWCNTs, drop coating, with (PANI-PMMA) blend and MB dye	0.0031	66.666	6.87 x 10^{-9}	1.45 x 10^{+8}
f-MWCNTs, drop coating	0.0009	10.231	3.22 x 10^{-8}	3.10 x 10^{+7}
f-MWCNTs, drop coating with (PANI-PMMA) blend and MB dye	0.0025	32.011	1.10 x 10^{-8}	9.06 x 10^{+7}
Short MWCNTs, drop coating	0.0013	3.923	4.399 x 10^{-8}	2.27 x 10^{+7}
Short MWCNTs, drop coating, with (PANI-PMMA) blend and MB dye	0.0028	34.576	9.047 x 10^{-9}	1.11 x 10^{+8}

Table 2: Some figures of Merit for the CNTs IR fabricated detectors prepared using n-type PSi substrate and worked at 5V forward bias voltage, Δf=1Hz, and active area A=1cm^2. The laser source of 10.6 μm wavelength.

Sample Conditions: (CNTs type, deposition technique, with or without blend coating layer)	R_λ [A/W]	G	NEP [Watt]	D* [(cm.Hz$^{1/2}$)/W]
MWCNTs, dip coating	0.0052	1.857	3.20×10^{-8}	$3.12 \times 10^{+7}$
MWCNTs, dip coating, with (PANI-PMMA) blend and MB dye	0.0095	48.879	1.41×10^{-12}	$7.08 \times 10^{+11}$
MWCNTs, drop coating	0.0050	4.045	2.20×10^{-8}	$4.53 \times 10^{+7}$
MWCNTs, drop coating, with (PANI-PMMA) blend and MB dye	0.0177	58.272	3.11×10^{-9}	$3.21 \times 10^{+8}$
f-MWCNTs, drop coating	0.0022	9.566	2.15×10^{-8}	$4.63 \times 10^{+7}$
f-MWCNTs, drop coating with (PANI-PMMA) blend and MB dye	0.0063	33.898	6.82×10^{-9}	$1.47 \times 10^{+8}$
Short MWCNTs, drop coating	0.0031	5.391	2.41×10^{-8}	$4.14 \times 10^{+7}$
Short MWCNTs, drop coating, with (PANI-PMMA) blend and MB dye	0.0051	40.854	6.93×10^{-9}	$1.44 \times 10^{+8}$

It can also noticed from these tables that the best results for R_λ, G, NEP, and D* were achieved for the detectors which were coated with the polymers blend and MB dye. The best sensitivity was obtained for the detector prepared using pure MWCNTs and dip-coating technique, which gave a gain of about 25 times the gain of the detector without the polymers blend and MB dye coating. While, for MWCNTs prepared by the drop-casting technique coated with the blend of (PANI-PMMA) polymers and MB dye, the gain became 14 times larger. The lowest value of NEP was 8.41×10^{-13} W with highest detectivity reaching $1.19 \times10^{+12}$ (cm.Hz$^{1/2}$)/W in the case of MWCNTs prepared by the dip-coating technique and operating at a forward bias voltage of 3V. All the obtained results are improved incomparably as compared with the results of [13] for the same conditions. These results are in agreement with the results of Saleh and Sadik [20, 21].

The goal of mixing PANI and PMMA polymers with MB dye of 1.56×10^{-4} M is to take advantage of their individual properties, such as the very smooth surface, transparency and sensitivity to dyes of the PMMA, and it is used as a host to PANI due to its good properties and good viscosity while PANI is not viscous. All these features result in a homogeneous mixture having a reasonable absorption beam around the 1000 cm^{-1} region, which is necessary to improve the detector performance in this region.

6. Conclusions

An infrared photoconductive detectors based on MWCNTs, working in the far-infrared region, and at room temperature, were successfully fabricated. They showed good photo-current response. The use of PSi led to an increase in the surface area as well as increase in the adhesion between the CNTs and the Si substrate, which resulted in good detector performance. The responsivity and the figures of merit of the photoconductive detector were improved by coating the MWCNTs films with a thin layer of polymers blend of (polyaniline- polymethyl methacrylate) polymers and methylene blue dye. The detectors with coated films of CNTs showed better performances , so this type of coating can be considered as a surface treatment of the detector film, which highly increased the responsivity and specific detectivity of the fabricated IR laser detector-based MWCNTs, which proved the role of the polymer on the enhancement of the performance of the IR photoconductive detectors. Coating the MWCNTs film with blend of (PANI-PMMA) polymers and MB dye is a good way to improve the performance and figures of merit of the photoconductive detector in the far-infrared region.

7. Acknowledgments

The authors express their thanks to Dr. Abdulkareem M. Ali, Dr. Mayson F. Al-Yas and Dr. Zaynab T. Al-Sheibani for their useful discussions.

References

[1] Rajeev Kumar, Mustaque A. Khan, A.V. Anupama, Saluru B. Krupanidhi, Balaram Sahoo, Infrared photodetectors based on multiwalled carbon nanotubes: Insights into the effect of nitrogen doping, J. Appl. Surface Science 538 (2021) 148187.

[2] Mohan Shankar, John B. Burchett, Human tracking systems using pyroelectric infrared detectors, Opt. Eng. 45 (2006) 106401:1-10.

[3] R. J. Keys, in Optical and Infrared Detectors, Topics in Applied Physics, second ed., Springer-Verlag, USA, 1980.

[4] E. Castro-Camus, L. Fu, J. Lloyd-Hughes, H. H. Tan, C. Jagadish, and M. B. Johnston, Photoconductive response correction for detectors of terahertz radiation, J. Appl. Phys. Amer. 104 (2008) 053113:1-7.

[5] G. Hasnain, B. F. Levine, Mid–infrared detectors in the 3- 5 μm band using bound to continuum state absorption in InGaAs/InAIAs multi-quantum well structures, Appl. Phys. Lett. 56 (1990) 770-772.

[6] M. E. Itkis, F. Borondics, A. Yu, and R. C. Haddon, Bolometric infrared photoresponse of suspended single-walled carbon nanotube films, Science. 312 (2006) 413–416.

[7] J. Zhang, N. Xi, K. Lai, Performance evaluation and analysis for carbon nanotube (CNT) based IR detectors, Proceeding of International Society for Optical Engineering, SPIE. 6542 (2007) 53-61.

[8] Standish, N. and Yu, A.B., Measurement of pore size distribution in ceramic, Powder Technology. 29 (1981) 151-156.

[9] M. S. Salem, M.J. Sailor, F.A. Harraz, T. Sakka, Y. H Ogata, Electrochemical stabilization of porous silicon multilayers for sensing various chemical compounds, J. Appl. Phys. 100 (2006) 083520:1-7.

[10] Shanxin Xiong, Xiangkai Zhang, Ru Wang, Yizhang Lu, Haifu Li, Jian Liu, Shuai Li, Zhu Qiu, Bohua Wu, Jia Chu, Xiaoqin Wang, Runlan Zhang, Ming Gong, Zhenming Chen, Preparation of covalently bonded polyaniline nanofibers/carbon nanotubes supercapacitor electrode materials using interfacial polymerization approach, J. Polym. Res. 26(4) (2019), doi.org/10.1007/s10965-019-1749-x.

[11] Jolly Bhadra1, Asma Alkareem1, and Noora Al-Thani1, A review of advances in the preparation and application of polyaniline based thermoset blends and composites, Journal of Polymer Research. 27:122 (2020), doi.org/10.1007/s10965-020-02052-1.

[12] Debasis Maity, Mathankumar Manoharan, and Ramasamy Thangavelu, Rajendra Kumar, Development of the PANI/MWCNT Nanocomposite-Based Fluorescent Sensor for Selective Detection of Aqueous Ammonia, American Chemical Society. 5 (2020) 8414–8422, doi.org/10.1021/acsomega.9b02885.

[13] Z. Ounaies, C. Park, K.E. Wise, E.J. Siochi and J.S. Harrison, Electrical properties of single-wall carbon nanotube reinforced polyimide, Composites Science and Technology. 63 (2003) 1637–1646.

[14] Ono, Y., Aoki T. and Ogasawara, T., Mechanical and electrical properties of carbon-nanotube composites, Proc. 48th Conference on structural strength in Japan, Kobe, (2006) 141-143.

[15] Maity, D.; Minitha, C. R.; Rajendra Kumar, R. T. Glucose Oxidase Immobilized Amine Terminated Multiwall Carbon Nanotubes/ Reduced Graphene Oxide/Polyaniline/Gold Nanoparticles Modified Screen-Printed Carbon Electrode for Highly Sensitive Amperometric Glucose Detection. Mater. Sci. Eng. C, 105 (2019) 110075.

[16] Wasan R. Saleh, Samar Y. Al-Dabagh, Marwa A. Al-Azzawi, Ghaida S. Muhammed, Abdulla M. Suhail, Far Infrared Photoconductive Detector Based on Multi-Wall Carbon Nanotubes, International Journal of Application or Innovation in Engineering & Management. 4 (2015) 62-67.

[17] Wasan R. Saleh, A carbon nanotubes photoconductive detector for middle and far-infrared regions based on porous silicon and a polyamide nylon polymer, Eur. Phys. J. Appl. Phys. 70 (2015) 30401:1-6, doi: 10.1051/epjap/2015150121.

[18] L. Liu and Y. Zhang, Multi-wall carbon nanotube as a new infrared detected material, Sensors and Actuators A. 116 (2004) 394-397.

[19] Asama N. Naje, Ola A. Noori, Mohammed A. Hamzah, Abdulla M. Suhail, Characterization of Carbon nanotubes Near- Infrared Photoconductive Detector, Journal of advances in physics. 9(1) (2015) 2318-2321.

[20] Wasan R. Saleh and Hawraa Sadik, Photodetection of Near and Middle IR Lasers by f-MWCNTs/ Polythiophen Nanocomposite Detector, Nano Hybrids and Composites, 18 (2017) 1-10, doi:10.4028/www.scientific.net/NHC.18.

[21] Hawraa Sadik, Wasan R. Sale, Naseer M. Hadi, Noon Kadhu, Near IR Photoconductive Detector Based on f-MWCNTs/Polythiophen Nanocomposite, Iraqi Journal of Science, 58 (2017) 868-877 doi:10.24996.ijs.2017.58.2B.11.

[22] Taqwa Y. Yousif and Asama N. Naje, Characterization of carbon nanotube decorated silver nanoparticles, J. Phys.: Conf. Ser. 1879: 032093 (2021), doi:10.1088/1742-6596/1879/3/032093.

[23] B.V. Mohan Kumar, Rajesh Thomas, Ambily Mathew, G. Mohan Rao, D. Mangalaraj1, N. Ponpandian1, C. Viswanathan1, Effect of catalyst concentration on the synthesis of MWCNT by single step pyrolysis, Advanced Materials Letters. 5(9) (2014) 543-548.

[24] P.G. Collins, M. Hersam, M. Arnold, R. Martel, Ph. Avouris, Current Saturation and Electrical Breakdown in Multiwalled Carbon Nanotubes, Phys. Rev. Lett. 86 (2001) 3128.

[25] Eman K. Hassan.2010. "Preparation and Development of Polymer Coated Zinc Oxide films for UV-Detection" M.Sc. Thesis, University of Baghdad, Dep. of Physics.

Nano Hybrids and Composites
ISSN: 2297-3370, Vol. 33, pp 105-132
© 2021 Trans Tech Publications Ltd, Switzerland

Submitted: 2020-10-21
Revised: 2021-08-17
Accepted: 2021-08-20
Online: 2021-10-11

Nonlinear Vibration and Stability Analysis of Functionally Graded Nanobeam Subjected to External Parametric Excitation and Thermal Load

Fateme Shayestenia[1,a], Mohadese Janmohammadi[1,b], Seyedabbas Sadatsakkak[1,c*] and Majid Ghadiri[1,d]

[1]Faculty of Engineering, Department of mechanics, Imam Khomeini International University, Qazvin, Iran, P.O. Box 34149-16818

[a]shayeste.f.h@gmail.com, [b]mohaddese.jmohammadi@gmail.com, [c]sakak@eng.ikiu.ac.ir, [d]ghadiri@eng.ikiu.ac.ir

*Corresponding author: sakak@eng.ikiu.ac.ir

Keywords: Nonlinear vibration; parametric excitation; instability region; multiple time scale method; thermal load; FG nanobeam

Abstract. Analysis of vibration stability of simply supported Euler-Bernoulli functionally graded (FG) nanobeam embedded in viscous elastic medium with thermal effect under external parametric excitation is presented in this work. An attempt has been made for the first time is investigating the effect of thermal load on dynamic behavior, amplitude response, instability region and bifurcation points of functionally graded nanobeam. Thermal loads are supposed to be uniform, linear or nonlinear distribution along the thickness direction. Nonlocal continuum theory and the principle of the minimum total potential energy are applied to derive the governing equations. The partial differential equations (PDE) are transported to the ordinary differential equations (ODE) by using the Petrov-Galerkin method and the multiple time scales method are manipulated to solve the motion equation. To study the effect of external parametric excitation and thermal effect, different temperature distributions along the thickness such as uniform, linear, and nonlinear distribution are considered. Moreover, stable and unstable regions and bifurcation points are determined. It is obtained that the thermal load can affect the amplitude response of FG nanobeam. Also, it is observed that the instability of the system is affected by the detuning parameter and the parametric excitation amplitude plays great role in the instability of system. Nanobeams are used in many devices like nanoresonators, nanosensors and nanoswitches. This paper is helpful for designing and manufacturing nanoscale structures specially nanoresonators under different thermal loads.

Introduction

Functionally graded materials, FGMs are new composite materials with amazing properties that are introduced by Japanese researchers[1]. As the temperature varies continuously along the thickness, the behavior of these materials changes exponentially. Because these materials play a substantial role in engineering, it is necessary to study and to investigate their mechanical and thermal behaviors. Functionally graded materials are using in industries such as aerospace, civil, plasma-facing biomaterial, marine, dental and orthopedic instrument, and sensors [2-5]. Functionally graded microbeam with changing temperature along thickness is presented by Jia et al. [6] to analyze buckling response under mechanical-thermal loads. The size-dependent static and dynamic behavior of functionally graded nanobeam embedded in elastic matrix is studied by utilizing Timoshenko nanobeam theory[7]. With the aid of high order shear deformation theory free vibration analysis of simply supported plate FG porous is studied. Material properties in FG plate vary across the thickness. Hamilton's principle is used to obtain the motion equation [8].

Because of the great influence of parametric excitation in dynamic behavior of the nanobeams, researchers are considering its impact on modeling and investigating nanostructures. Huang et al. [9] presented instability of nanobeam subjected to parametric excitation. By employing nonlocal elasticity theory, an exact method to study the dynamic instability was introduced. Wang et al.[10]emphasize the influence of the external parametric excitation on nanobeam. Parametric stabilization of electrostatically actuated micro-scale cantilever beam is investigate by Krylov et al.[11]. Negative and positive bifurcation points can be changed by applying parametric excitation. Parametric excitation is utilized to investigate the oscillations of composite plates[12]. A double-walled nanobeam subjected to parametric external load is investigated by Wang[13]. He presented the relationship between the amplitude and frequency. By using nonlocal continuum theory, the stability, steady-state response, and natural frequency of nanobeam under various forces are revealed by li et al. [14]. Jump phenomena and frequency response of a structure can be affected by electric voltage and shock force pulse. They intensify vibration properties in order to generate resonance[15].

The classical continuum theory neglects the size effect, so other theories like the couple stress theory, the strain gradient theory, and the nonlocal continuum theory are introduced in order to predict the vibrational properties of nanobeam [16]. The static bending and free vibration of piezoelectric nanobeam is studied to realize the effect of nanoscale and surface energy. Eringen's nonlocal elasticity theory is employed to demonstrate the long-range atom interactions. The outcome of this research is that in nanoscale as the thickness reduces the effect of surface stresses become more considerable[17]. Ansari and his co-authors presented the Timoshenko beam which is subjected to the magnetic force and external electric voltage with assuming the effect of temperature by employing nonlocal elasticity theory[18]. Highly nonlinear integro-differential equation governing electrically actuated nanobeams made of functionally graded material is presented. In the framework of Euler-Bernoulli nanobeam theory, the modified couple stress theory and Gurtin-Murdoch surface elasticity are employed to take account the size effects of nanoscale structures[19]. Eringen's nonlocal elasticity model of the nanotube is established to inquire the effect of size on the bifurcation characteristic of fluid-conveying nanobeam [20]. Free vibration and resonance of frequencies of nano-resonator is investigated by Eltaher et al.[21]. The employ Timoshenko and Euler-Bernoulli theories to take account thick and thin beams. Nano-scale size dependency of nano-resonator is considered by manipulating Eringen's nonlocal differential theory. this model can be used in designing and producing nanoresonators in nanoelectromechanical systems (NEMS). Vibrational characteristics of functionally graded (FG) cracked microbeam embedded in elastic foundation and subjected to the thermal and magnetic loads are studied. Present work is concentrated on temperature dependent material properties and the effect of small scale parameter. By employing Euler-Bernoulli beam theory and Hamilton's principle the motion equation is derived[22]. Post-buckling and free vibration of geometrically imperfect multilayer nanobeam is investigated. The small-size effect is model according to the nonlocal elasticity differential model[23]. By utilizing nonlocal strain gradient theory, the oscillation properties of the nanotube with fractional-order derivative damping are studied. The results indicate that fundamental frequency and time response are impressing by the fractional-order derivative damping [24]. Bending characteristics of perforated microbeams subjected to various loading pattern are investigated. In order to consider the effect of nonclassical size dependency, the modified couple stress theory is employed[25]. Buckling and post-buckling of single-walled carbon nanotube(SWCNT) rested on nonlinear elastic matrix is investigated by considering the effect of size dependency[26]. The sandwich functionally graded nanoplate embedded in changing Winkler elastic matrix is presented to inspect the bending deflection and stress distribution in the basis of new quasi 3D hyperbolic shear theory in conjoint with nonlocal strain gradient theory[27]. To investigate static analysis of simply supported cross-ply carbon nanotubes reinforced composite subjected to different loading conditions is presented. The nonlocal strain gradient constitutive relation is used to evaluate the size-dependence of nano-scale[28]. In the framework of the Euler-Bernoulli beam theory, the stiffness-softening and stiffness-hardening size effect of the nanotube is investigated via nonlocal geometric theory. In this way, nonlinear bending and post-buckling are analyzed and investigated. The novel numerical procedure is introduced to predict nonlinear buckling and postbuckling stability of clamped-clamped single walled carbon nanotube (SWCNT) surrounded by nonlinear elastic matrix. Nanoscale effect of carbon nanotubes (CNTs) is include by using energy-equivalent model[29].

Vibration behavior is affected due to temperature, and therefore it is necessary to consider the thermal loads. Many researchers who study and investigate in the vibration field assumed thermal environment. Vibrational properties of the nanotube with the magneto-electro-thermo-elastic field are researched by Ebrahimi and Barati. They realized that the vibration characteristics of functionally graded nanobeam are affected by these parameters [30]. Functionally graded sandwich beam is presented in order to investigate to thermal buckling. The equation of stability is obtained by employing generalized higher-order shear deformation beam theory[31]. Different transformation method is utilized to model the Euler-Bernoulli nanobeam under thermal field [32]. Buckling and vibrational treatment of functionally graded nanotube with regarding to thermal impact are introduced by Zheng [33] to the exposure uncertainty model. Cracked nanobeam with the elastic foundation via finite element approach is alleged to study and analyze the thermal transverse oscillations. In this study, the Petrov-Galerkin technique is employed to solve the equation of motion and determine the natural frequency [34]. Nonlinear thermal buckling and post-buckling of symmetry angle-ply laminated composite beam is studied. The motion equation is derived on the basis of Reddy's higher order shear deformation plate theory and von Karman strain-displacement relation[35]. Buckling analysis of nanobeam subjected to diverse thermal and magnetic and electric force based on the Euler-Bernoulli beam theory is studied by Alibeigi and her coworkers [36]. In the purported model, if the temperature reduces or increases instantly, the temperature will distribute along the cross-section uniformly[37].

With the respect to developmental works on nonlinear vibration, an attempt has been made for the first time is investigating and analyzing the influence of thermal load on vibrational characteristics and instability regions. It is presented that the thermal load and external parametric excitation can change the position of bifurcation points and instability of FG nanobeam. The temperature has linear, nonlinear, and uniform distribution along the thickness. Different power-low exponents are supposed to investigate dynamic behavior. By the nonlocal continuum beam model and Hamilton's principle based on minimum potential energy the governing equation of FG nanobeam is obtained and to solve the equation, the perturbation method of multiple time scale is employed. The different kinds of bifurcation points are determined and the effect of thermal load on bifurcation points is presented. Finally, to investigate the instability of the system the trivial and nontrivial solutions are discussed.

Problem Formulation

Fig. 1 indicates that the FG nanotube resting on the viscous elastic medium with length L and diameter d. The Euler-Bernoulli nanobeam is exposed to the axial parametric excitation with the harmonic frequency Ω. The deformation is along the z-axis with the displacement defined by w.

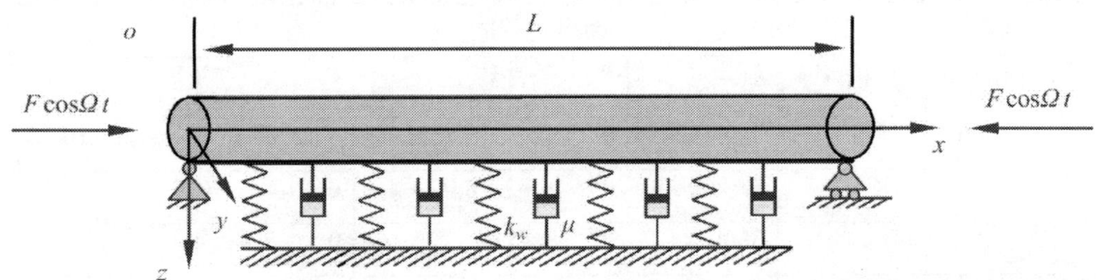

Fig. 1. The Euler-Bernoulli FG nanobeam embedded in the viscous elastic foundation under axial load.

The displacements have the following forms:

$$u_x(x, z, t) = u(x, t) - z\frac{\partial w}{\partial x}, \quad u_y = 0, \quad u_z(x, z, t) = w(x, t). \tag{1}$$

Where w is transverse displacement and u is the axial displacement. The nanobeam is made by composing two different materials. The materials properties of the beam are assumed to vary continuously in the thickness direction (z-axis direction) according to the rule of the mixture as:

$$P_f = P_C V_c + P_M V_m. \tag{2}$$

Where P_C and P_M are the material properties at the upper (ceramic) and lower (metal) surfaces and V defines the volume fraction, respectively, and subscripts M and C refer to the metal and ceramic constituents, respectively. A simple power-law is considered to describe the variation of material properties from pure metal at the bottom surface (Z=-$h/2$) to pure ceramic at the top surface (Z=$h/2$) of the beam as:

$$V_C + V_M = 1. \tag{3a}$$

$$V_C = (\frac{z}{h} + \frac{1}{2})^p \quad ; \quad V_M = 1 - (\frac{z}{h} + \frac{1}{2})^p. \tag{3b}$$

Where p is the power-law exponent. Effective material properties of the FG beam such as mass density (ρ), thermal expansion coefficient (a) and Young's modulus (E) can be determined by substituting Eq. (4) into Eq. (2) as:

$$\rho(z) = (\rho_C - \rho_M)\left(\frac{z}{h} + \frac{1}{2}\right)^p + \rho_M. \tag{4a}$$

$$a(z) = (a_C - a_M)\left(\frac{z}{h} + \frac{1}{2}\right)^p + a_M \quad . \tag{4b}$$

$$E(z) = (E_C - E_M)\left(\frac{z}{h} + \frac{1}{2}\right)^p + E_M \quad . \tag{4c}$$

Since the material properties are dependent on temperature, it is essential to consider the following nonlinear equation of material thermo-elastic properties to predict the behavior of functionally graded nanobeam due to temperature as [38]:

$$P = P_0(P_{-1}T^{-1} + 1 + P_1 T + P_2 T^2 + P_3 T^3). \tag{5}$$

Where P_0, P_{-1}, P_1, P_2 and P_3 are the temperature coefficients that are shown in Table 1 for Al_2O_3 and $SUS304$.

Table 1. Temperature dependant coefficients of Young's modulus, thermal expansion, and mass density [39].

Material	Properties	P₀	P₋₁	P₁	P₂	P₃
Al₂O₃	E[Pa]	349.55e+9	0	−3.853e−4	4.027e−7	1.673e−10
	a[K⁻¹]	6.8269e-6	0	1.838e−4	0	0
	ρ[Kg/m³]	3800	0	0	0	0
SUS304	E[Pa]	201.04e+9	0	3.079e−4	−6.534e−7	0
	a[K⁻¹]	12.330e−6	0	8.086e−4	0	0
	ρ[Kg/m³]	8166	0	0	0	0

For the nonlinear vibration, the nonzero von Karman nonlinear strain should be considered as:

$$\varepsilon = \varepsilon_0 + \varepsilon_1. \tag{6a}$$

$$\varepsilon_0 = \frac{\partial u}{\partial x} + \frac{1}{2}\left(\frac{\partial w}{\partial x}\right)^2, \quad \varepsilon_1 = -z\kappa, \quad \kappa = -\frac{\partial^2 w}{\partial x^2}. \tag{6b}$$

Where k and ε_0 are bending strain and nonlinear extensional strain, respectively. Then, the von Karman strain is:

$$\varepsilon_{non} = \frac{\partial u}{\partial x} + \frac{1}{2}\left(\frac{\partial w}{\partial x}\right)^2 - z\frac{\partial^2 w}{\partial x^2}. \tag{7}$$

By using the Hamilton's principle[40], the governing equation of transverse vibration can be derived as:

$$\int_0^t \delta(T - U + W)dt = 0. \tag{8}$$

Where W denotes the work done by external forces, U and T denote the strain and kinetic energies, respectively.

The variation of the strain energy can be expressed as:

$$\delta U = \int_0^L (N(\delta\varepsilon_{xx}^0) - M(\delta k^0))dx. \tag{9}$$

In which N and M are the axial force and bending moment, respectively.

$$N = \int_A \sigma_x \, dA, \qquad M = \int_A z\sigma_x \, dA. \tag{10}$$

The kinetic energy for the Euler-Bernoulli nanobeam is given by:

$$T = \frac{1}{2}\int_0^L \int_A \rho(z,T)\left(\left(\frac{\partial u_x}{\partial t}\right)^2 + \left(\frac{\partial u_z}{\partial t}\right)^2\right)dAdx. \tag{11}$$

$$\delta T = \int_0^L I_0 \left(\frac{\partial u}{\partial t}\frac{\partial \delta u}{\partial t} + \frac{\partial w}{\partial t}\frac{\partial \delta w}{\partial t}\right)dx. \tag{12}$$

Where $I_0 = \int_A \rho(z,T)\, dA$ is the mass moment of inertia.

The variation of external forces can be expressed as:

$$\delta W = -\int_0^L[\frac{1}{2}F\,\cos\Omega t(\frac{\partial w}{\partial x})^2\delta w + \frac{1}{2}N^T\,(\frac{\partial w}{\partial x})^2\delta w + q\delta w]dx. \tag{13}$$

The transverse load $q(x, t)$ can be defined as:

$$q = k_w w + c_d \frac{\partial w}{\partial t}. \tag{14}$$

Where k_w and c_d are linear Winkler and viscoelastic damping parameters of the foundation, respectively.

N^T is thermal resultant and can be assumed as:

$$N^T = \int_A \frac{E(z,T)}{1-v^2}a(z,T)(T - T_0)dz. \tag{15}$$

For the functionally graded nanobeam which the beam thickness is thin enough, the temperature distribution is supposed to be changed uniformly, linearly, and nonlinearly along the thickness as follows:

$$T(z) = T_L - \Delta T(\frac{z}{h} + \frac{1}{2})^c. \tag{16}$$

Where $\Delta T = T_U - T_L$ and c is the temperature index. By setting $c = 0$, $c = 1, c = 2$ the uniform, linear and nonlinear rise of temperature, respectively, can be obtained.

Substituting Eqs. (9), (12), and (13) into Eq. (8), Euler-Lagrange equations of transverse motion can be obtained as:

$$\frac{\partial N}{\partial x} = 0. \tag{17a}$$

$$\frac{\partial^2 M}{\partial x^2} + \frac{\partial}{\partial x}\left(N \frac{\partial w}{\partial x}\right) - (F \cos \Omega t + N^T)\frac{\partial^2 w}{\partial x^2} + c_d \frac{\partial w}{\partial t} + k_w w = I_0 \frac{\partial^2 w}{\partial t^2}. \tag{17b}$$

Based on Eringen's elasticity theory [41, 42], the stress at a reference point x in a body is assumed as a function of strains of all points in the near region. The stress tensor $\sigma_{ij}(x)$ and strain tensor ε_{ij} can be obtained as:

$$\sigma_{ij}(x) = \int \tilde{\alpha}(|x - \acute{x}|, \tau)C_{ijkl}\varepsilon_{kl}(\acute{x})d\bar{V}(\acute{x}) \tag{18a}$$

$$\varepsilon_{ij} = \frac{1}{2}(u_{i,j} + u_{j,i}) \tag{18b}$$

Where, $\tilde{\alpha}(|x - \acute{x}|, \tau)$, C_{ijkl}, $|x - \acute{x}|$, $\tilde{\alpha}$ and u_i represent the nonlocal kernel, elastic modulus, the Euclidean distance, nonlocal modulus and displacement vector, respectively. Based on Ref. [43], the equivalent differential constitutive equation can be formed as:

$$(1 - (e_0. a)^2 \nabla^2)\sigma = C:\varepsilon \tag{18c}$$

Here, ∇^2, C and $e_0. a$ demonstrate the Laplacian operator, fourth-order elasticity tensor, and the small-scale parameter, respectively. Where, e_0 is the material constant that is determined experimentally or by other ways. a is an internal characteristic (e.g., lattice parameter). The nonlocal stress-strain relation can be obtained as:

$$\sigma_{xx} - (e_0. a)^2 \frac{\partial^2 \sigma_{xx}}{\partial x^2} = E\varepsilon_{xx} \tag{18d}$$

Moment-curvature relation can be established as bellow:

$$M - (e_0. a)^2 \frac{\partial^2 M}{\partial x^2} = EI \frac{\partial^2 w}{\partial x^2} \tag{18e}$$

$$N_{xx} = \frac{A_{xx}}{2L} \int_0^L (\frac{\partial w}{\partial x})^2 dx - \frac{B_{xx}}{L}[\frac{\partial w}{\partial x}(L, t) - \frac{\partial w}{\partial x}(0, t)]. \tag{19}$$

The explicit relation for the nonlocal axial force from Eqs. (19) And (17a), [3, 4, 44] can be written as:

$$N = \frac{A_{xx}}{2L} \int_0^L (\frac{\partial w}{\partial x})^2 dx - \frac{B_{xx}}{L}\left[\frac{\partial w}{\partial x}(L, t) - \frac{\partial w}{\partial x}(0, t)\right]. \tag{20}$$

Also, the explicit relation for the nonlocal bending moment from Eqs. (19) and (17a) can be derived as:

$$M = B_{xx}\left[\frac{\partial u}{\partial x} + \frac{1}{2}\left(\frac{\partial w}{\partial x}\right)^2\right] - C_{xx}\frac{\partial^2 w}{\partial x^2}.\tag{21}$$

In which the cross-sectional rigidities are defined as follows:

$$(A_{xx}, B_{xx}, C_{xx}) = \int_A E(z,T)(1, z, z^2)\, dA.\tag{22}$$

That B_{xx} is negligible in comparison with other components (A_{xx} and C_{xx}). [45]
Substituting Eq. (20) and Eq. (21), into Eq. (17b), the nonlinear vibration equation of nonlocal Euler- Bernoulli FG nanobeam can be derived as:

$$I_0\frac{\partial^2 w}{\partial t^2} + C_{xx}\frac{\partial^4 w}{\partial^4 x} + \left[F\cos\Omega t + N^T - \frac{A_{xx}}{2L}\int_0^L\left(\frac{\partial w}{\partial x}\right)^2 dx\right]\frac{\partial^2 w}{\partial^2 x} + k_w w + c_d\frac{\partial w}{\partial t} = 0.\tag{23}$$

Galerkin Technique

To convert a continuous operator problem to a discrete problem, the Galerkin technique is an excellent choice. To transfer the partial differential equation (PDE) to the ordinary differential equation (ODE), the Galerkin method is utilized[46]. The flexural displacement can be expressed as:

$$w(x,t) = \varphi(t)\sin\frac{k\pi}{L}.\tag{24}$$

Substituting Eq. (24) into Eq. (23), the governing equation of motion can be obtained as follows:

$$B_1\frac{\partial^2\varphi}{\partial t^2} + B_2\frac{\partial\varphi}{\partial t} + B_3\varphi + B_4\varphi^3 = 0.\tag{25}$$

B_1, B_2, B_3 and B_4 can be express as:

$$B_1 = I_0.\tag{26a}$$

$$B_2 = \mu.\tag{26b}$$

$$B_3 = k_w - F\cos\Omega t\left(\frac{k\pi}{L}\right)^2 - N^T\left(\frac{k\pi}{L}\right)^2 + C_{xx}\left(\frac{k\pi}{L}\right)^4.\tag{26c}$$

$$B_4 = \frac{A_{xx}}{4}\left(\frac{k\pi}{L}\right)^4.\tag{26d}$$

The following normalized dimensional variables are presented as bellow:

$$\tau = \frac{\Omega}{2}t, \quad \varphi = r\psi, \quad r = \sqrt{\frac{I}{A}}.\tag{27}$$

Then, it can be obtained as:

$$\ddot{\psi} + S^2\psi = -\varepsilon\eta_1\dot{\psi} - \varepsilon\eta_2\psi^3 + \varepsilon\eta_3\psi\cos2\tau.\tag{28}$$

$$S^2 = \frac{4\omega_0^2}{\Omega^2}.\tag{29a}$$

$$\omega_0 = \sqrt{\frac{1}{I_0}\left[k_w - N^T\left(\frac{k\pi}{L}\right)^2 + C_{xx}\left(\frac{k\pi}{L}\right)^4\right]}. \tag{29b}$$

$$\eta_1 = \frac{2\mu}{I_0\Omega^2}. \tag{29c}$$

$$\eta_2 = \frac{A_{xx}\left(\frac{k\pi}{L}\right)^4 I}{A\Omega^2 I_0}. \tag{29d}$$

$$\eta_3 = \frac{4F}{I_0\Omega^2}\left(\frac{k\pi}{L}\right)^2. \tag{29e}$$

For the principal parametric resonance case of the nanobeam, the non-dimensional frequency can be given as:

$$S = 1 + \varepsilon\sigma. \tag{30}$$

The multiple time scale method which was presented by Nayfeh and Mook [47, 48] for the first time, has been manipulate to solve Eq.(28).

For fundamental frequency (S=1) and set of approximate solution can be define as follow:

$$\psi(\tau,\varepsilon) = \psi_0(T_0,T_1) + \varepsilon\psi_1(T_0,T_1) + \cdots. \tag{31}$$

Now, two time scales are supposed as bellow:

$$T_0 = \tau, \quad T_1 = \varepsilon\tau. \tag{32}$$

Where, T_0 and T_1 define the fast time scale characterizing the motions and the slow time characterizing the modulation of the phases and amplitude, respectively, and ε is the scaling parameter. It is mentioned that T_1 is presented the nonlinear part of the set of approximation. In accordance with Eq. (32), it can be obtained as follows:

$$\frac{d}{dT} = \frac{d}{dT_0} + \varepsilon\frac{d}{dT_1} + \cdots = D_0 + \varepsilon D_1 + \cdots. \tag{33a}$$

$$\frac{d^2}{dT^2} = \frac{d^2}{dT_0^2} + 2\varepsilon\frac{d}{dT_0}\frac{d}{dT_1} + \cdots = D_0^2 + 2\varepsilon D_0 D_1 + \cdots. \tag{33b}$$

By Substituting Eq. (31) into Eq. (28), the coefficients of ε^0 and ε^1 are equated with zero, thus, it can be obtained as follows:

$$\varepsilon^0: D_0^2\psi_0 + \omega_0^2\psi_0. \tag{34a}$$

$$\varepsilon^1: D_0^2\psi_1 + \psi_1 = -2D_0D_1\psi_0 - 2\sigma\psi_0 - \eta_2\psi_0^2 - \eta_1D_0\psi_0 + \eta_3\psi_0\cos 2T_0. \tag{34b}$$

The solution of Eq. (34a) can be reached as:

$$\psi_0 = A(T_1)\exp(iT_0) + \overline{A}(T_1)\exp(iT_0). \tag{35}$$

A demonstrates an unknown complex function and $\overline{A}$ shows the complex conjugate of A, respectively. Using equation (35) and substituting in equation (34b), it can be obtained as follows:

$$D_0{}^2\psi_1 + \psi_1 = [-2i\acute{A} - 2\sigma A - 3\eta_2 A^2\bar{A} - i\eta_1 A + \tfrac{\eta_3}{2}\bar{A}]\exp(iT_0) - \eta_2 A^3 \exp(3iT_0) +$$
$$\tfrac{\eta_3}{2}A\exp(3iT_0) + c.c. \tag{36}$$

Where $i=\sqrt{-1}$ and c.c. is the complex conjugate of the preceding terms. In the next step, it is needed to eliminate the secular terms in the equation.

$$-2i\acute{A}(T_1) - 2\sigma A(T_1) - 3\eta_2 A^2\bar{A} - i\eta_1 A(T_1) + \tfrac{\eta_3}{2}\bar{A}(T_1). \tag{37}$$

It is needed to express a polar form for A as follow:

$$A(T_1) = \tfrac{1}{2}\alpha(T_1)\exp\left[i\beta(T_1)\right]. \tag{38}$$

Where $\alpha(T_1)$ and $\beta(T_1)$ are real functions. Substituting Eq. (38) into Eq. (37) and separating the real and imaginary part, it can be derived:

$$\acute{\alpha} = -\tfrac{1}{2}\eta_1\alpha - \tfrac{\eta_3}{4}\alpha\sin(2\beta). \tag{39a}$$

$$\alpha\acute{\beta} = \sigma\alpha + \tfrac{3\eta_2}{8}\alpha^3 - \tfrac{\eta_3}{4}\alpha\cos(2\beta). \tag{39b}$$

furthermore, the steady-state response can be obtained by supposing $\acute{\alpha} = 0$ and $\alpha\acute{\beta}=0$, thus, modulation equations for the principal parametric resonance are define as:

$$-\tfrac{1}{2}\eta_1 - \left(\tfrac{\eta_3}{4}\right)\sin(2\beta) = 0. \tag{40a}$$

$$\sigma + \tfrac{3\eta_2}{8}\alpha^2 - \tfrac{\eta_3}{4}\cos(2\beta). \tag{40b}$$

Trivial steady-state response

To represent the stability of the system, the Cartesian form is denoted as $(A=\tfrac{1}{2}(p - iq)e^{i\beta T_1})$, where p and q are the function of T_1 and substituting A in Eq. (37) as below:

$$i(\acute{p} - i\acute{q}) - \beta(p - iq) + i\eta_1(p - iq) + \tfrac{3\eta_2}{8}(p^2 - q^2 - 2ipq)(p + iq) + \tfrac{\eta_3}{4}(p + iq) = 0. \tag{41}$$

To make the equation of system autonomous, it is assumed that $\beta=\tfrac{\sigma}{2}$ and by substituting it in Eq. (41) it can be obtained as:

$$\acute{p} = -\tfrac{\sigma}{2}q - \eta_1 p + \tfrac{3\eta_2}{8}q(p^2 + q^2) - \tfrac{\eta_3}{4}q. \tag{42a}$$

$$\acute{q} = \tfrac{\sigma}{2}p - \eta_1 q - \tfrac{3\eta_2}{8}p(p^2 + q^2) - \tfrac{\eta_3}{4}p. \tag{42b}$$

In order to study the stability of the system for the trivial steady-state response, it is supposed $p=q=0$, and the Jacobian matrix of Eq. (42a) and Eq. (42b) are represented as following:

$$A = \begin{bmatrix} \dfrac{d\acute{p}}{dp} & \dfrac{d\acute{p}}{dq} \\ \dfrac{d\acute{q}}{dp} & \dfrac{d\acute{q}}{dq} \end{bmatrix} = \begin{bmatrix} -\eta_1 & -\dfrac{\sigma}{2} - \dfrac{\eta_3}{4} \\ \dfrac{\sigma}{2} - \dfrac{\eta_3}{4} & -\eta_1 \end{bmatrix}.$$

The determinant and trace of matrix A are specified as bellow:

$$\Delta = \eta_1{}^2 + (\frac{\sigma^2}{4} - \frac{\eta_3{}^2}{16}).$$

$$\tau = -2\eta_1.$$

By assuming $\eta_1 > 0$ and $K = \frac{\eta_3}{2\omega_0}$, it is possible to analyse the stability of the system.

Non-trivial steady-state response

To investigate the stability of the FG nanobeam, the determination of the matrix should reset to zero. In non-trivial steady-state solution $\alpha \neq 0$. So it can be obtained as bellow:

$$-\frac{1}{2}\eta_1 - \left(\frac{\eta_3}{4}\right)\sin(2\beta) = 0. \tag{43a}$$

$$\sigma + \frac{3\eta_2}{8}\alpha^2 - \frac{\eta_3}{4}\cos(2\beta). \tag{43b}$$

In Eq. (43a) and Eq. (43b), the trigonometric function $(sin^2\beta + cos^2\beta = 1)$ can be utilized, so it can be expressed as bellow:

$$\left[(\tfrac{1}{2}\eta_1)^2 + (\sigma + \tfrac{3\eta_2}{8}\alpha^2)^2\right] = (\tfrac{\eta_3}{4})^2. \tag{44}$$

$$(\sigma + \tfrac{3\eta_2}{8}\alpha^2)^2 = (\tfrac{\eta_3}{4})^2 - \frac{\eta_1{}^2}{4}. \tag{45}$$

By considering that $\alpha = 0$, the detuning parameter for the positive bifurcation point is defined as bellow:

$$\sigma = +\sqrt{(\tfrac{\eta_3}{4})^2 - \frac{\eta_1{}^2}{4}}. \tag{46}$$

Results and Discussion

In this section, the numerical results are presented and discussed in detail. The Euler-Bernoulli FG nanobeam exposed to the external parametric excitation is investigated and discussed. Moreover, the thermal field is considered. It is an important step to verify the numerical results. To achieve this aim, the results are compared with available literature to be sure about the accuracy of the present formulation.

Validation of the study

To evaluate the correctness of the present results, it is proper to contrast the results of the present study with the results of available researches. In order to do that, eliminating the thermal effect, the present results can be compared with the results reported by Wang et al. [8]. Fig. 2.can provide a good comparison between the present results and the results reported by Wang et al. [8]. This figure indicates the relation between the response of nanobeam and detuning parameter. A good agreement can also be observed from the results.

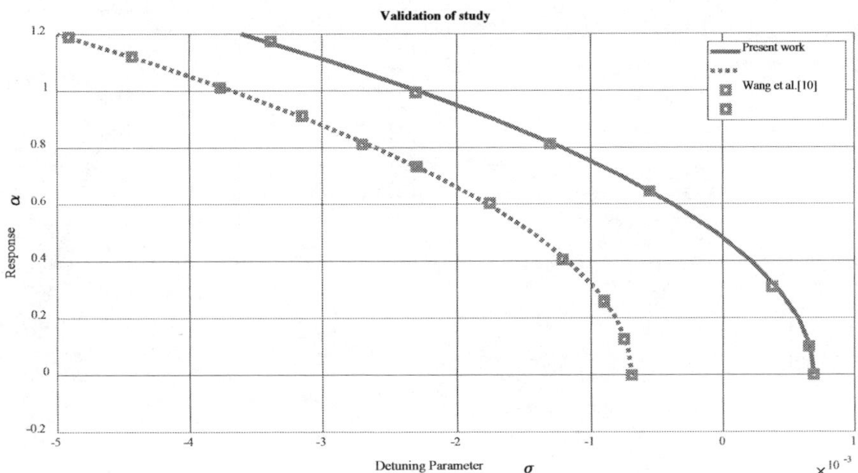

Fig. 2. The relation between amplitude response and detuning parameter of Euler-Bernoulli FG nanobeam compared with the results reported by Wang et al. [10]

Numerical results and discussion

Now, numerical results on the vibration behavior of FG Euler-Bernoulli nanobeam embedded in the viscous elastic medium by applying external parametric excitation and thermal load, are investigated and disserted. In this section, concentrating on the effect of thermal load and parametric excitation, the relation between the detuning parameter and amplitude response and bifurcation points are studied. Plots and graphs for different parameters are presented to make a better understanding.

The effect of applying thermal load is shown in Figs. *3-6*. In these plots, the relation between the temperature index and the amplitude response is presented. We will have different states. It can be observed that when the frequency of vibration is larger than the natural frequency (detuning parameter is positive), with the increase in the temperature index, the amplitude response decreases, and when the frequency of vibration is smaller than the natural frequency (detuning parameter is negative), with the increase in temperature, the amplitude response increases, too. It means that thermal load can influence the amplitude response and vibrational behavior.

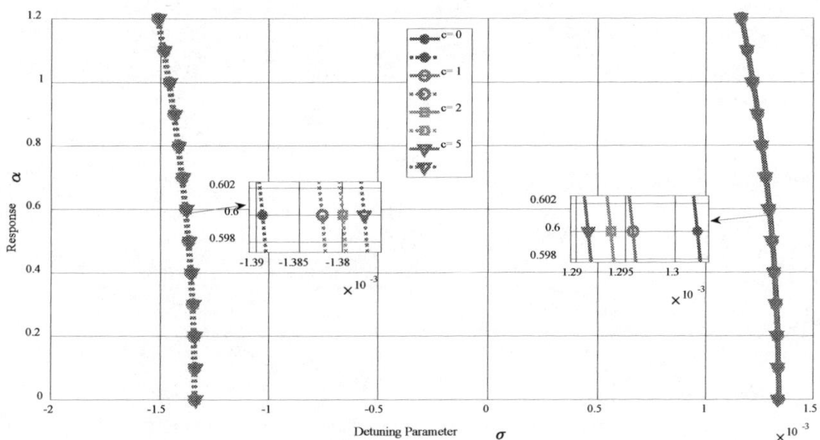

Fig. 3. The relation between the temperature index and amplitude response ($P=0$).

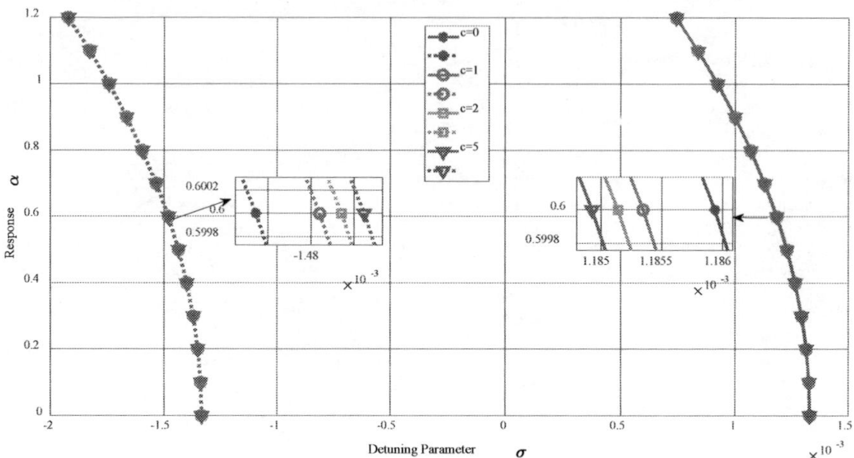

Fig. 4. The relation between the temperature index and amplitude response ($P=1$).

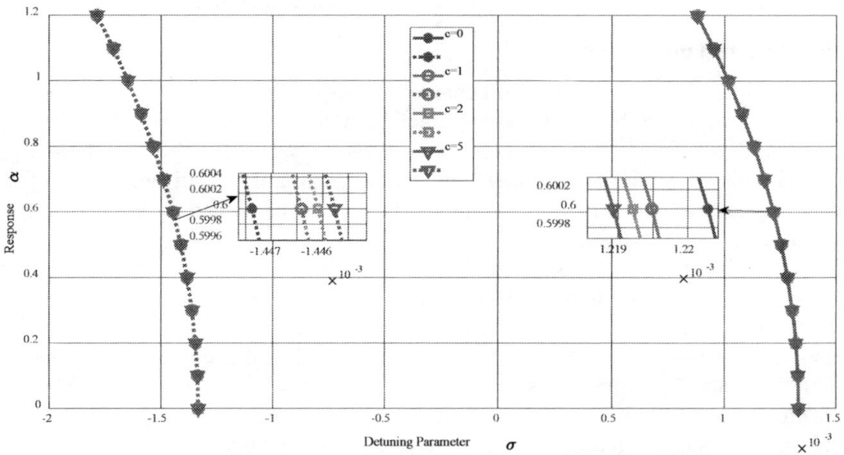

Fig. 5. The relation between the temperature index and amplitude response ($P=2$).

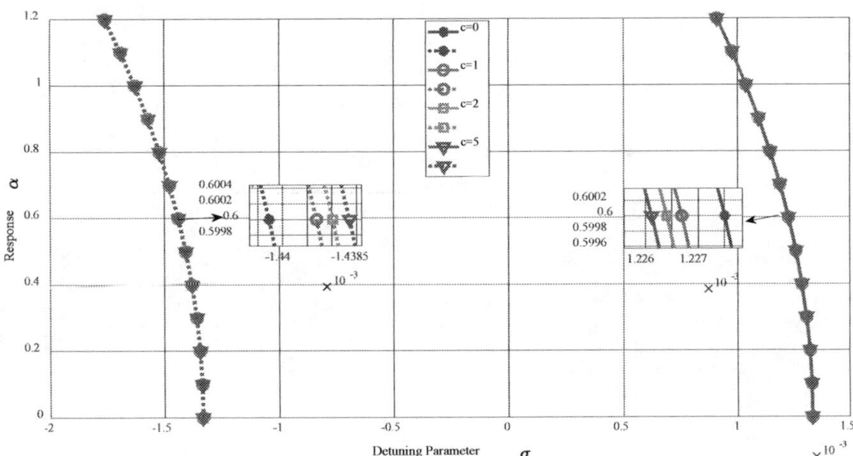

Fig. 6. The relation between the temperature index and amplitude response ($P=4$).

To evaluate the effect of variation of material properties from pure metal at bottom surface to pure ceramic at the top surface, different power-law exponents are applied. Figs. *7-10* demonstrate that although employing the various power-law exponents do not change the instability of FG nanobeam and stable and unstable regions, but with the increase in power-law exponents, the amplitude response grows. The amplitude response of simply-supported nanobeam is affected by power-law exponent and variation in material leads to change in the growth of amplitude response.

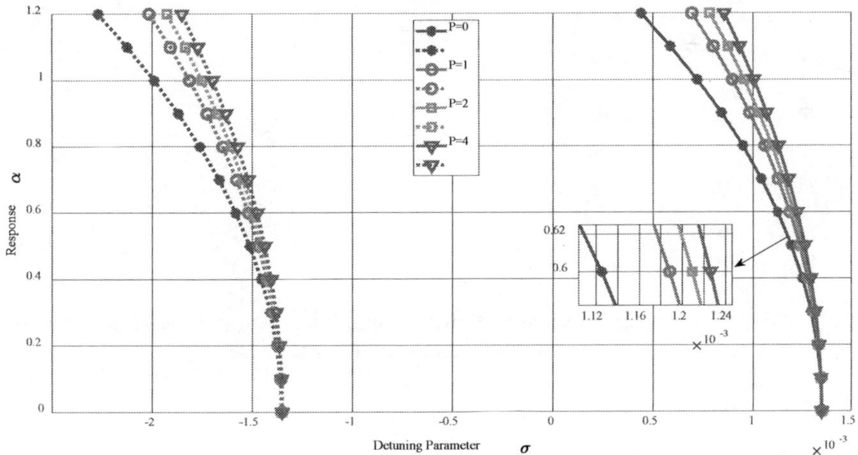

Fig.7. The effect of detuning parameter against the amplitude response applying the different power-law exponent ($c=0$).

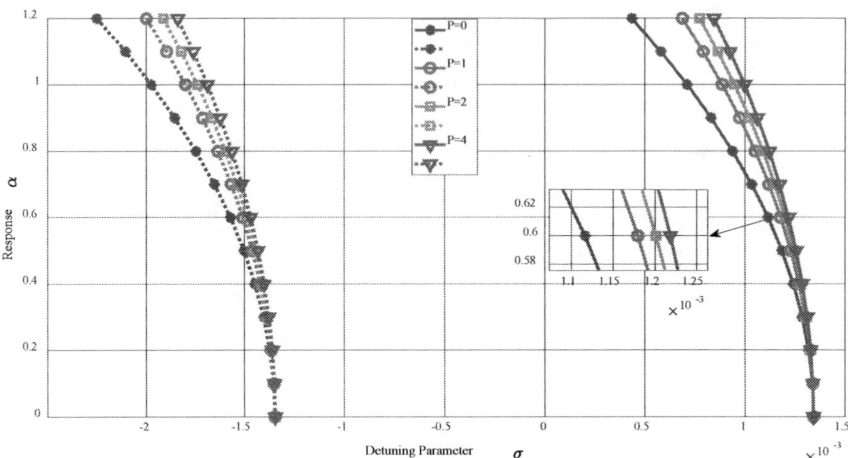

Fig. 8. The effect of detuning parameter against the amplitude response applying the different power-law exponent ($c=1$).

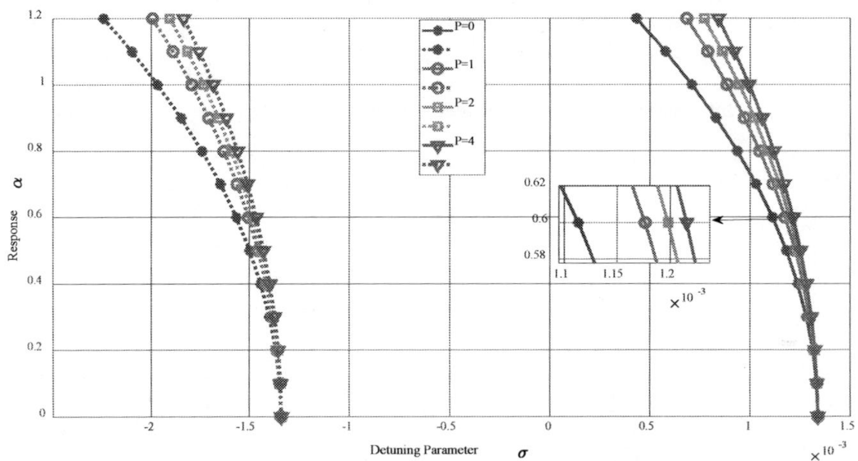

Fig. 9. The effect of detuning parameter against the amplitude response applying the different power-law exponent ($c=2$).

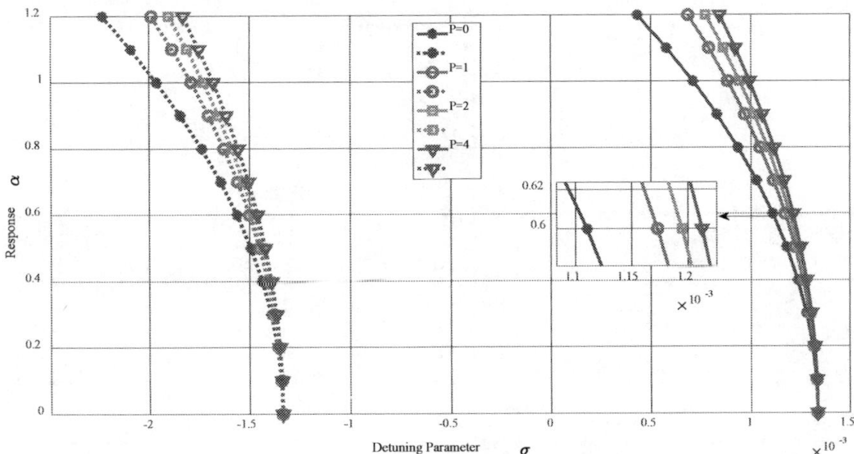

Fig. 10. The effect of detuning parameter against the amplitude response applying the different power-law exponent ($c=5$).

Various axial parametric excitations are assumed as $F=0.0005EA$, $0.001EA$ and $0.002EA$, respectively. Figs. *11-14* show that the parametric excitation amplitude can affect the instability of FG nanobeam. The parametric excitation amplitude increases, the curves move far from each other and eventually, the stable region becomes larger. It can be realized that the parametric excitation amplitude have considerable impact on the stability of the system. Increasing the amplitude of parametric excitation enlarges the stable region. Also, it can be realized that the growth of amplitude response is not affected by the amplitude of parametric excitation.

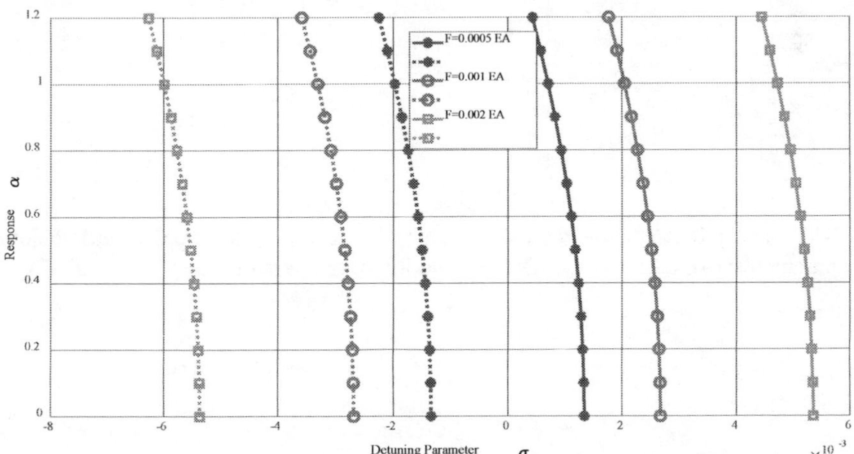

Fig. 11. The relation between the detuning parameter and amplitude response with the effect of parametric excitation. ($c=1$. The distribution of temperature is linear and $P=0$).

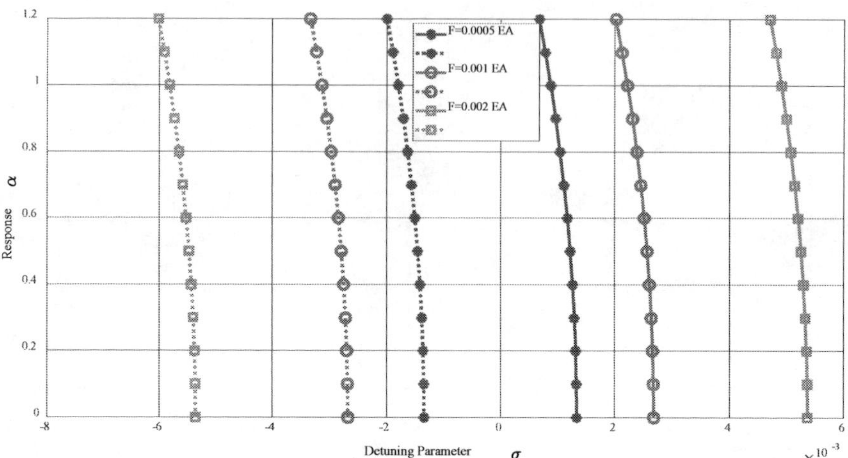

Fig. 12. The relation between the detuning parameter and amplitude response with the effect of parametric excitation. ($c=1$. The distribution of temperature is linear and $P=1$).

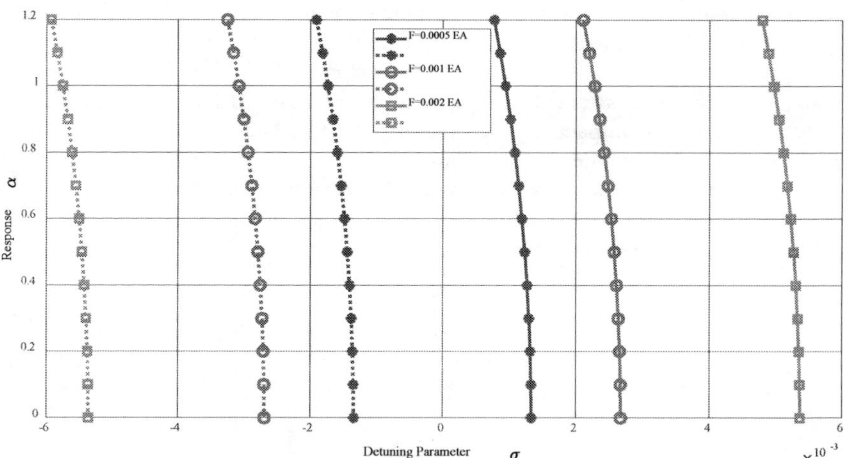

Fig. 13. The relation between the detuning parameter and amplitude response with the effect of parametric excitation. ($c=1$. The distribution of temperature is linear and $P=2$)

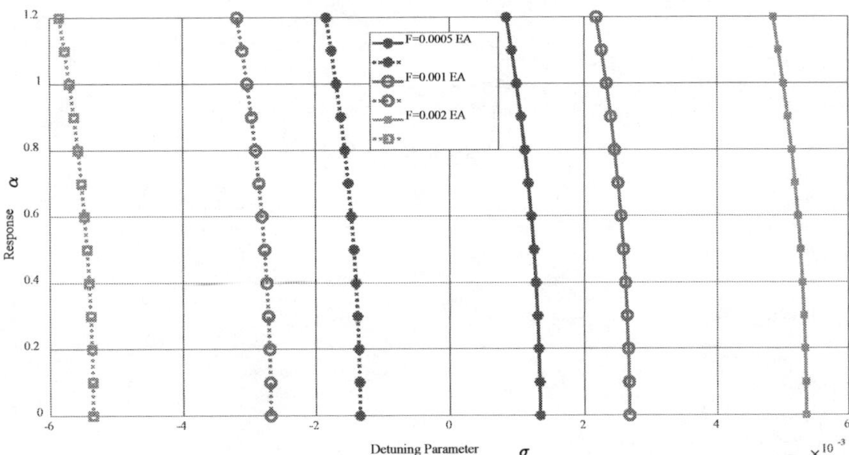

Fig. 14. The relation between the detuning parameter and amplitude response with the effect of parametric excitation. ($c=1$. The distribution of temperature is linear and $P=4$).

Figs. *15-18* indicate the role of matrix stiffness on the amplitude response. The distance between stable and unstable solutions becomes smaller as the matrix stiffness increases. Matrix stiffness (Winkler coefficient) can shift the bifurcation points and affect the instability. The instability region gets smaller with an increase in the Winkler coefficient. As shown in Figs. *15-18* change in Winkler coefficient is not able to influence on the growth of amplitude response.

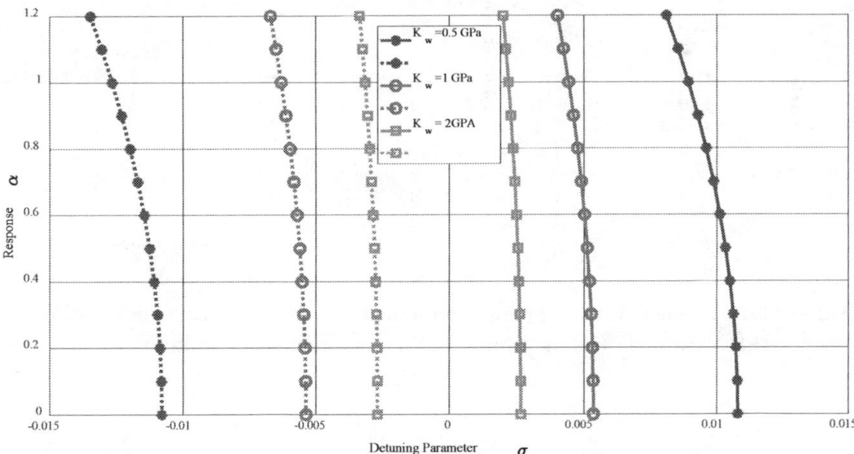

Fig. 15. The relation between detuning parameter and amplitude response with the effect of matrix stiffness. (c=1. The distribution of temperature is linear and P=0).

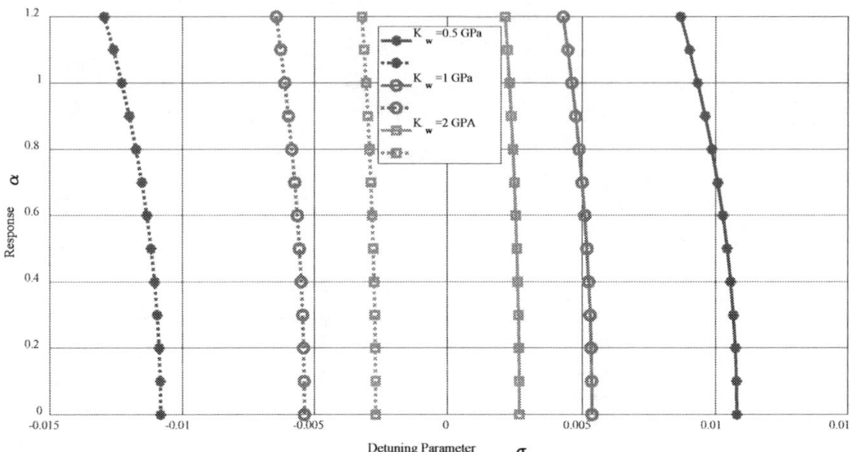

Fig. 16. The relation between detuning parameter and amplitude response with the effect of matrix stiffness. (c=1. The distribution of temperature is linear and P=1).

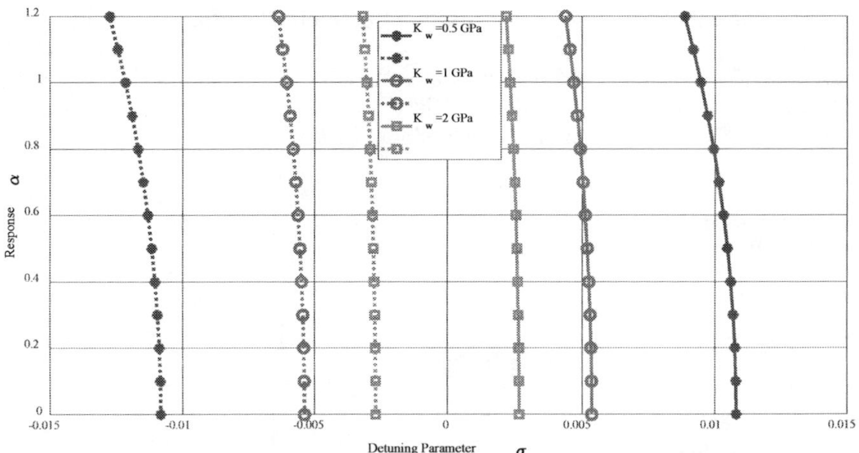

Fig. 17. The relation between detuning parameter and amplitude response with the effect of matrix stiffness. (c=1. The distribution of temperature is linear and P=2).

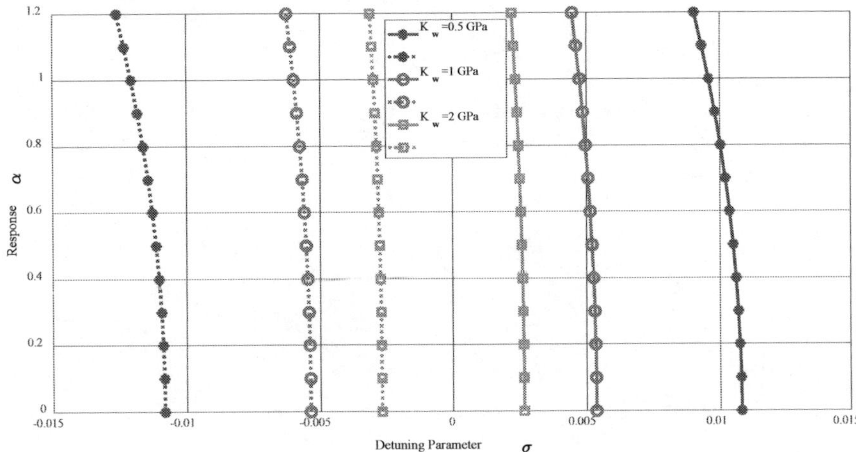

Fig. 18. The relation between detuning parameter and amplitude response with the effect of matrix stiffness. (c=1. The distribution of temperature is linear and P=4)

The temperature changes along with the thickness of the Euler-Bernoulli nanobeam. The growth of amplitude response can be affected by applying different temperatures for the top surface of the nanobeam. In order to investigate the effect of changing the temperature to the top surface on vibrational behaviour and instability regions, in Figs. *19-22* we assume that the distribution of temperature is linear and power-law exponents change also, in Figs. *23-25* power-law exponent is held constant while the distribution of temperature changes. Figs. *19-25* show that when the detuning parameter is positive, with increasing temperature of the top surface, the amplitude response increases too and while the detuning parameter is negative, this behaviour changed. It means that with increasing temperature, the amplitude response decrease. Also, it can be observed that changing the temperature is not able to change the instability of the system. The position of bifurcation points does not change. Supercritical and supercritical pitchfork bifurcations are not affected by thermal load.

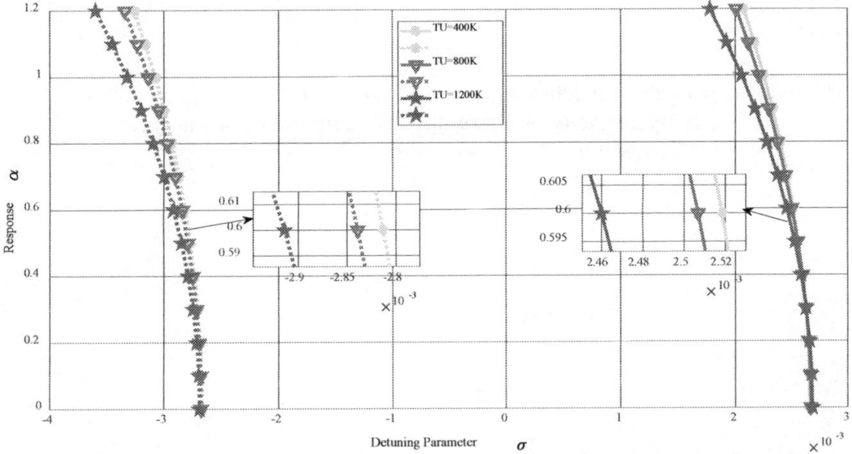

Fig. 19. The relation between the detuning parameter and amplitude response for various value of top surface temperature (c=1 distribution of temperature is linear and P=0).

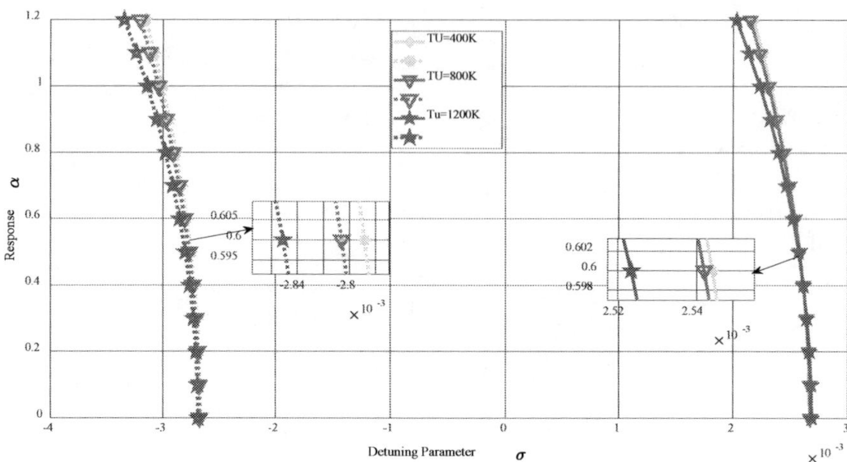

Fig. 20. The relation between the detuning parameter and amplitude response for various value of top surface temperature (c=1 distribution of temperature is linear and P=1).

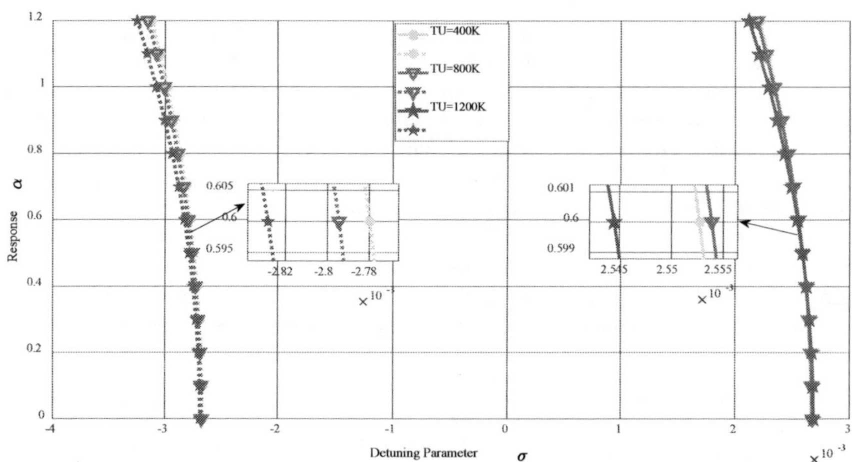

Fig. 21. The relation between the detuning parameter and amplitude response for various value of top surface temperature (c=1 distribution of temperature is linear and P=2).

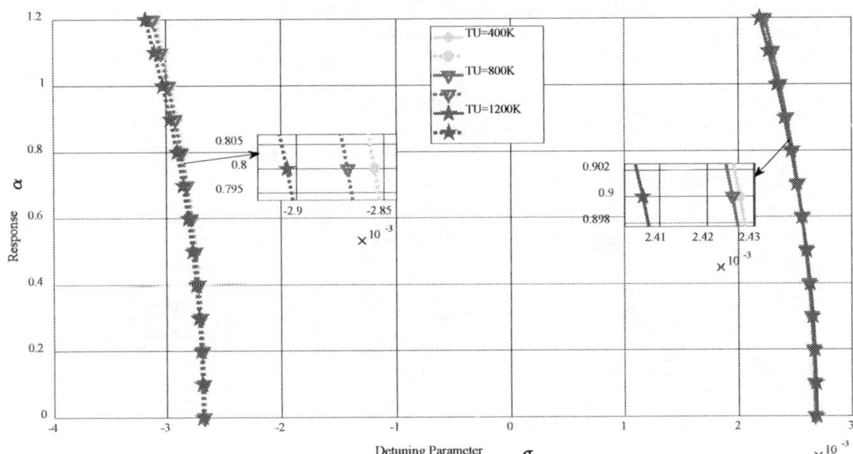

Fig. 22. The relation between the detuning parameter and amplitude response for various value of top surface temperature (c=1 distribution of temperature is linear and P=4).

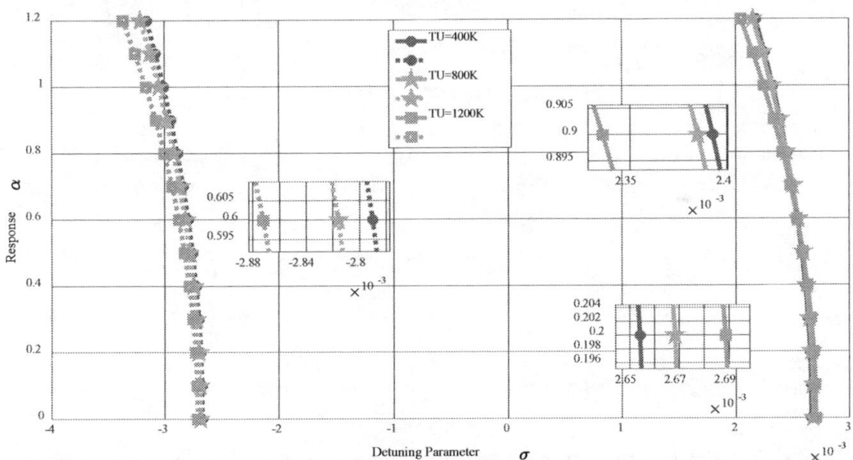

Fig. 23. The relation between the detuning parameter and amplitude response for different value of top surface temperature (c=0).

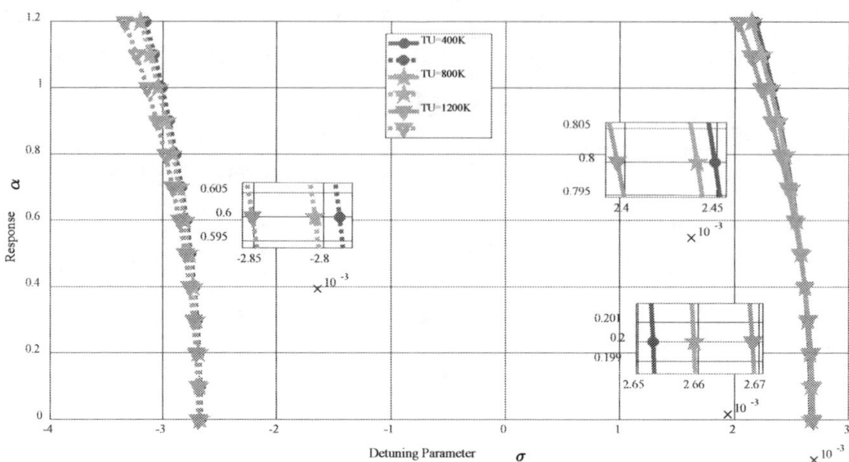

Fig. 24. The relation between the detuning parameter and amplitude response for different value of top surface temperature (c=1).

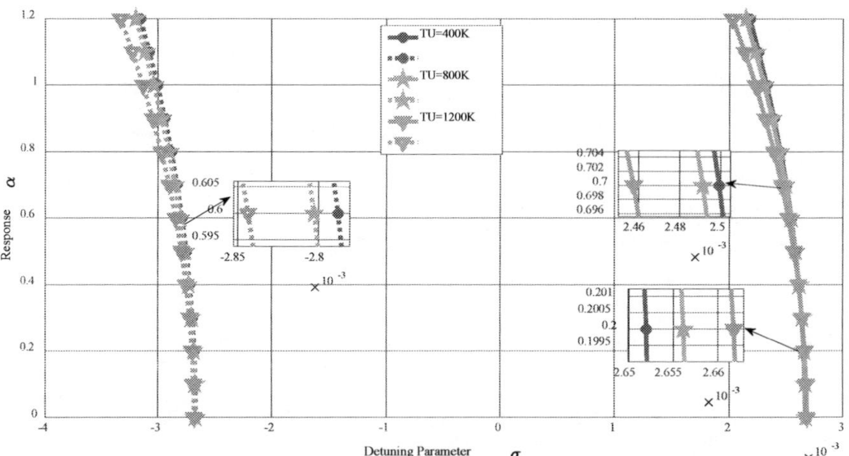

Fig. 25. The relation between the detuning parameter and amplitude response for different value of top surface temperature (c=2).

The thermal load applied on a nanobeam leads to make material expand or contract. The rise in temperature causes a reduction in terms of stiffness and strength. This phenomenon tends to increase in amplitude response of nanobeam when subjected to a higher temperature. The behaviour of nanobeam under thermal load can be described by these definitions.

Fig. 26 indicates the influence of the detuning parameter with versos the amplitude of parametric excitation. The region (ı) is always stable and this region is relevant to the trivial solution. In the region (ıı) two solutions exist. One trivial solution is unstable and another non-trivial solution is stable. There are three solutions in the region (ııı) comprising one trivial solution (stable) and two nontrivial solutions (stable and unstable).

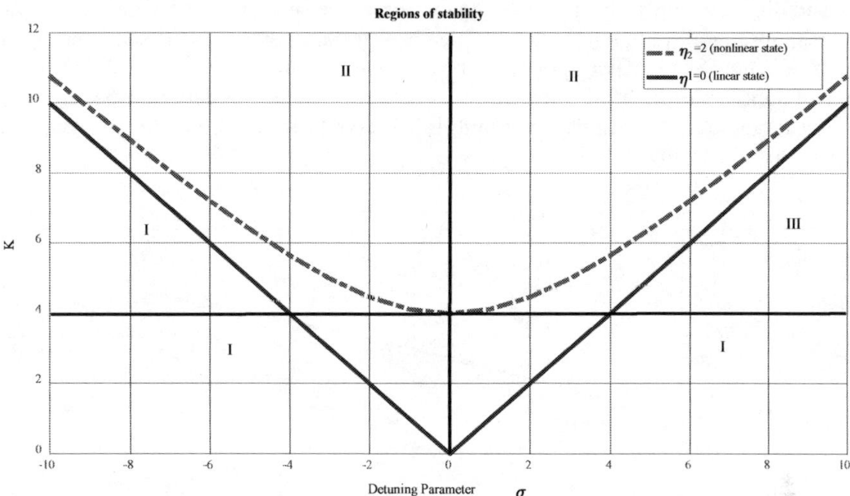

Fig. 26. The relation between the detuning parameter and the force amplitude of external parametric excitation for linear and nonlinear states in different regions of stability(I, II, III).

Fig. *27* shows that with the increase in detuning parameter, different bifurcation points can be observed. The point (σ_1) shows the subcritical pitchfork bifurcation and the point (σ_2) indicates the supercritical pitchfork bifurcations. The instability of the system varies in the point (σ_1) and also in the point (σ_2). It is necessary to imply that to achieve these results, there are two situation, the amplitude of the axial force and non-linearity should be $(K > 2\eta_1)$ and $(\eta_2 > 0)$, respectively.

In Fig. *27* the detuning parameter is raised while the force amplitude is taken constant. Before the point (σ_1) trivial solution is stable and realized solution is obtained by Eq. (45). Between the point (σ_1) and the point (σ_2) trivial solution is unstable and again realized solution is obtained by Eq. (45). After the point (σ_2) only a trivial solution is observed and it is stable. It is presented that the detuning parameter has a significant role in the instability of systems.

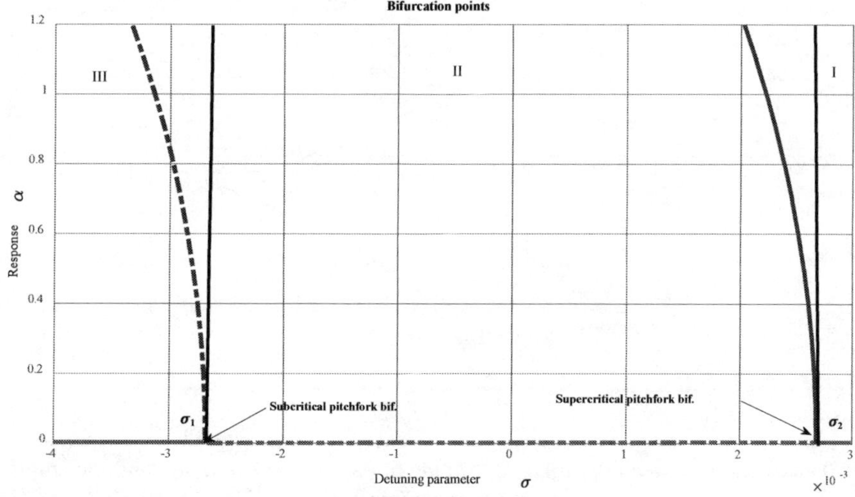

Fig. 27. The detuning parameter versus the response amplitude.

Fig. *28* and Fig. *29* imply the relationship between the amplitude of external excitation (K) against the response amplitude (α). With the increase of the detuning parameter, the instability of the system varies, too. Supercritical pitchfork bifurcation is achieved when the detuning parameter is positive and subcritical pitchfork bifurcation is achieved when the detuning parameter becomes negative. Also, saddle-node bifurcation point is observed. In saddle-node, the manner of FG nanobeam varies and unstable solution changes to a stable solution.

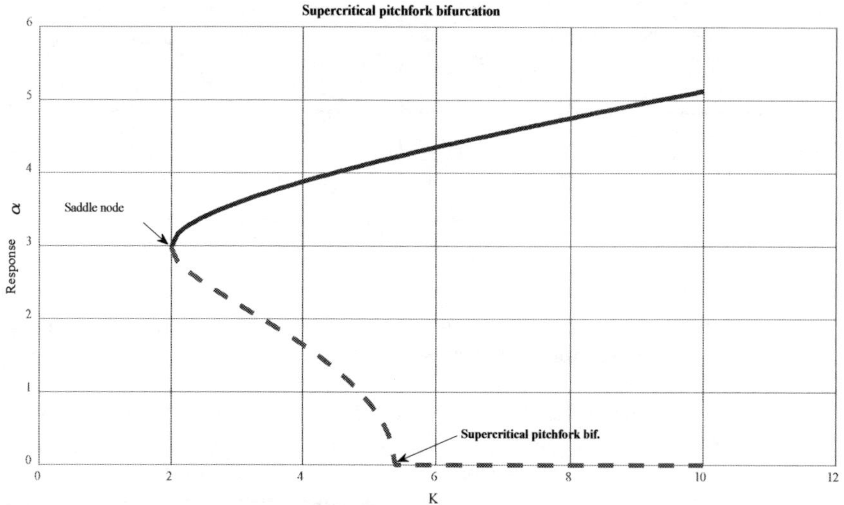

Fig. 28. The relation between the axial force amplitude of parametric excitation and amplitude response in positive σ.

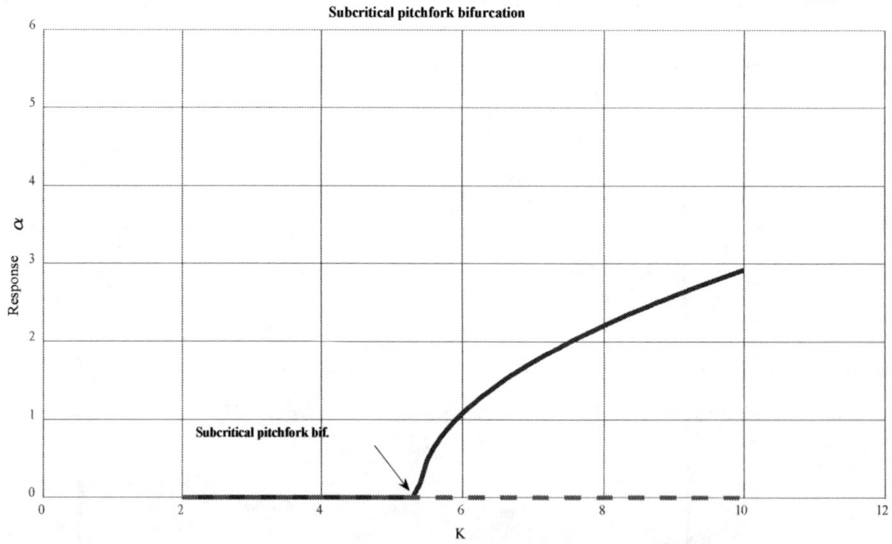

Fig. 29. The relation between the force amplitude of parametric excitation and the amplitude response in negative σ.

Conclusions

The vibration stability of the FG Euler-Bernoulli nanobeam exposed to external axial force and thermal load is the main object of the present research. For the first time, the temperature distribution is assumed to be constant, linear and nonlinear distribution along the thickness. In the next step, the nonlocal continuum theory, the nonlinear Von Karman strain-displacement relation, and Hamilton's principle are employed to derive the nonlinear governing equation. Then, for solving the governing equation, the Galerkin method and multiple time scale are used. Trivial and nontrivial solutions are presented to study instability of functionally graded nanobeam. Different bifurcation points are determined in order to investigate vibrational characteristics of FG nanobeam. The results show that external excitation loads can alter the dynamic instability of FG nanobeam. Uniform, linear and nonlinear distribution of temperature changes the magnitude of amplitude response. The stability of the system does not change by applying thermal load. The gap between bifurcation points can be influenced by the Winkler coefficient. The major outcomes of this paper can be mentioned as:

- The growth of amplitude response is affected by the thermal load.

- Subcritical and supercritical pitchfork bifurcation points and instability of the system are not affected by thermal load.

- Increasing the parametric excitation amplitude, enlarges the gap between stable and unstable regions.

- Various power-law exponents do not change the instability of FG nanobeam, but with the increase in power-law exponents, the amplitude response grows.

- Winkler coefficient can affect the instability of the system. The gap between bifurcation points enlarges with reduction of Winkler coefficient.

- Subcritical and supercritical pitchfork bifurcations depend on the sign of detuning parameter.

References

[1] M. Yamanouchi, M. Koizumi, T. Hirai, I. Shiota, Proc. First Int. Symp. on Functionally Gradient Materials, FGM Forum Tokyo, Japan, 1990.

[2] F. Ebrahimi, E. Salari, Thermo-mechanical vibration analysis of nonlocal temperature-dependent FG nanobeams with various boundary conditions, Composites Part B: Engineering 78 (2015) 272-290.

[3] A. Fallah, M. Aghdam, Nonlinear free vibration and post-buckling analysis of functionally graded beams on nonlinear elastic foundation, European Journal of Mechanics-A/Solids 30(4) (2011) 571-583.

[4] A. Fallah, M. Aghdam, Thermo-mechanical buckling and nonlinear free vibration analysis of functionally graded beams on nonlinear elastic foundation, Composites Part B: Engineering 43(3) (2012) 1523-1530.

[5] E. Shahabinejad, N. Shafiei, M. Ghadiri, Influence of Temperature Change on Modal Analysis of Rotary Functionally Graded Nano-beam in Thermal Environment, Journal of Solid Mechanics Vol 10(4) (2018) 779-803.

[6] X. Jia, L. Ke, X. Zhong, Y. Sun, J. Yang, S. Kitipornchai, Thermal-mechanical-electrical buckling behavior of functionally graded micro-beams based on modified couple stress theory, Composite Structures 202 (2018) 625-634.

[7] R.A. Shanab, M.A. Attia, S.A. Mohamed, N.A. Mohamed, Effect of microstructure and surface energy on the static and dynamic characteristics of FG Timoshenko nanobeam embedded in an elastic medium, Journal of Nano Research, Trans Tech Publ, 2020, pp. 97-117.

[8] S. Merdaci, Free Vibration Analysis of Composite Material Plates" Case of a Typical Functionally Graded FG Plates Ceramic/Metal" with Porosities, Nano Hybrids and Composites, Trans Tech Publ, 2019, pp. 69-83.

[9] Y. Huang, J. Fu, A. Liu, Dynamic instability of Euler–Bernoulli nanobeams subject to parametric excitation, Composites Part B: Engineering 164 (2019) 226-234.

[10] Y.-Z. Wang, Y.-S. Wang, L.-L. Ke, Nonlinear vibration of carbon nanotube embedded in viscous elastic matrix under parametric excitation by nonlocal continuum theory, Physica E: Low-dimensional Systems and Nanostructures 83 (2016) 195-200.

[11] S. Krylov, I. Harari, Y. Cohen, Stabilization of electrostatically actuated microstructures using parametric excitation, Journal of Micromechanics and Microengineering 15(6) (2005) 1188.

[12] M. Darabi, R. Ganesan, Non-linear vibration and dynamic instability of internally-thickness-tapered composite plates under parametric excitation, Composite Structures 176 (2017) 82-104.

[13] Y.-Z. Wang, Nonlinear internal resonance of double-walled nanobeams under parametric excitation by nonlocal continuum theory, Applied Mathematical Modelling 48 (2017) 621-634.

[14] C. Li, C.W. Lim, J. Yu, Dynamics and stability of transverse vibrations of nonlocal nanobeams with a variable axial load, Smart Materials and Structures 20(1) (2010) 015023.

[15] M. Ghadiri, S.H. S Hosseini, Nonlinear forced vibration of graphene/piezoelectric sandwich nanoplates subjected to a mechanical shock, Journal of Sandwich Structures & Materials (2019) 1099636219849647.

[16] A.C. Eringen, Nonlocal polar elastic continua, International journal of engineering science 10(1) (1972) 1-16.

[17] M.A. Eltaher, F.-A. Omar, W.S. Abdalla, A.M. Kabeel, A.E. Alshorbagy, Mechanical analysis of cutout piezoelectric nonlocal nanobeam including surface energy effects, Structural Engineering and Mechanics 76(1) (2020) 141-151.

[18] R. Ansari, R. Gholami, H. Rouhi, Size-dependent nonlinear forced vibration analysis of magneto-electro-thermo-elastic Timoshenko nanobeams based upon the nonlocal elasticity theory, Composite Structures 126 (2015) 216-226.

[19] R.M. Abo-Bakr, M.A. Eltaher, M.A. Attia, Pull-in and freestanding instability of actuated functionally graded nanobeams including surface and stiffening effects, Engineering with Computers (2020) 1-22.

[20] A. Farajpour, M.H. Ghayesh, H. Farokhi, A coupled nonlinear continuum model for bifurcation behaviour of fluid-conveying nanotubes incorporating internal energy loss, Microfluidics and Nanofluidics 23(3) (2019) 34.

[21] M.A. Eltaher, N.A. Mohamed, Vibration of nonlocal perforated nanobeams with general boundary conditions, Smart Structures and Systems 25(4) (2020) 501-514.

[22] I. Esen, C. Özarpa, M.A. Eltaher, Free Vibration of a Cracked FG Microbeam Embedded in an Elastic Matrix and Exposed to Magnetic Field in a Thermal Environment, Composite Structures 113552.

[23] S.A. Emam, M.A. Eltaher, M.E. Khater, W.S. Abdalla, Postbuckling and free vibration of multilayer imperfect nanobeams under a pre-stress load, Applied Sciences 8(11) (2018) 2238.

[24] A.K. Jha, S.S. Dasgupta, Mathematical modeling of a fractionally damped nonlinear nanobeam via nonlocal continuum approach, Proceedings of the Institution of Mechanical Engineers, Part C: Journal of Mechanical Engineering Science (2019) 0954406219866467.

[25] A.A. Abdelrahman, H.E. Abd-El-Mottaleb, M.A. Eltaher, On bending analysis of perforated microbeams including the microstructure effects, Structural Engineering and Mechanics 76(6) (2020) 765.

[26] N. Mohamed, S. Mohamed, M. Eltaher, Buckling and post-buckling behaviors of higher order carbon nanotubes using energy-equivalent model, Engineering with Computers (2020) 1-14.

[27] A.A. Daikh, M.S.A. Houari, M.A. Eltaher, A novel nonlocal strain gradient Quasi-3D bending analysis of sigmoid functionally graded sandwich nanoplates, Composite Structures (2020) 113347.

[28] A.A. Daikh, A. Drai, M.S.A. Houari, M.A. Eltaher, Static analysis of multilayer nonlocal strain gradient nanobeam reinforced by carbon nanotubes, Steel and Composite Structures 36(6) (2020) 643-656.

[29] M.A. Eltaher, N. Mohamed, S. Mohamed, L.F. Seddek, Postbuckling of curved carbon nanotubes using energy equivalent model, Journal of Nano Research, Trans Tech Publ, 2019, pp. 136-157.

[30] F. Ebrahimi, M.R. Barati, Vibration analysis of smart piezoelectrically actuated nanobeams subjected to magneto-electrical field in thermal environment, Journal of Vibration and Control 24(3) (2018) 549-564.

[31] A.A. Daikh, M. Guerroudj, M. El Adjrami, A. Megueni, Thermal buckling of functionally graded sandwich beams, Advanced Materials Research, Trans Tech Publ, 2020, pp. 43-59.

[32] R. Hamza-Cherif, M. Meradjah, M. Zidour, A. Tounsi, S. Belmahi, T. Bensattalah, Vibration analysis of nano beam using differential transform method including thermal effect, Journal of Nano Research, Trans Tech Publ, 2018, pp. 1-14.

[33] Z. Lv, H. Liu, Uncertainty modeling for vibration and buckling behaviors of functionally graded nanobeams in thermal environment, Composite Structures 184 (2018) 1165-1176.

[34] A. Aria, M. Friswell, T. Rabczuk, Thermal vibration analysis of cracked nanobeams embedded in an elastic matrix using finite element analysis, Composite Structures 212 (2019) 118-128.

[35] X.P. Chang, Z. Liang, Q.Y. Liu, Buckling and post-buckling analysis of symmetrically angle-ply laminated composite beams under thermal environments, Advanced Materials Research, Trans Tech Publ, 2011, pp. 182-186.

[36] B. Alibeigi, Y.T. Beni, F. Mehralian, On the thermal buckling of magneto-electro-elastic piezoelectric nanobeams, The European Physical Journal Plus 133(3) (2018) 133.

[37] Y. Zhao, C. Huang, Temperature Effects on Nonlinear Vibration Behaviors of Euler-Bernoulli Beams with Different Boundary Conditions, Shock and Vibration 2018 (2018).

[38] Y.S. Touloukian, Thermophysical properties of high temperature solid materials: elements, Macmillan1967.

[39] Y.S. Touloukian, Thermophysical properties of high temperature solid materials, Macmillan 1967.

[40] S.S. Rao, Vibration of continuous systems, Wiley Online Library2007.

[41] A.C. Eringen, Theories of nonlocal plasticity, International Journal of Engineering Science 21(7) (1983) 741-751.

[42] A.C. Eringen, D. Edelen, On nonlocal elasticity, International journal of engineering science 10(3) (1972) 233-248.

[43] A.C. Eringen, On differential equations of nonlocal elasticity and solutions of screw dislocation and surface waves, Journal of applied physics 54(9) (1983) 4703-4710.

[44] S.A. Emam, A.H. Nayfeh, Postbuckling and free vibrations of composite beams, Composite Structures 88(4) (2009) 636-642.

[45] M. Eltaher, A. Alshorbagy, F. Mahmoud, Determination of neutral axis position and its effect on natural frequencies of functionally graded macro/nanobeams, Composite Structures 99 (2013) 193-201.

[46] B.A. Finlayson, The method of weighted residuals and variational principles, SIAM2013.

[47] A.H. Nayfeh, Introduction to perturbation techniques.-New York, Willey & Sons (1981).

[48] A.H. Nayfeh, D.T. Mook, Nonlinear oscillations, John Wiley & Sons2008.

Keyword Index

A

Advanced Materials 61
AFM 47
Attenuated Total Reflection 73

B

Biopolymer.................................... 35

C

Carbon Nanotubes......................... 93
Composite 1
Composite Material 13

D

DFT 35
Differential Scanning Calorimetry..... 73
Diffusion 61
DS Model 61
Dynamic 83

E

ESP 35
Exfoliation..................................... 47

F

FG Nanobeam 105
Fg-Nanoplate................................. 83
Finite Element Method.................. 13
FTIR 35

G

Glass Transition Temperature........ 73
Graphene.................................. 47, 61

H

HOMO-LUMO 35

I

Implants... 13
Infrared Detector 93
Instability Region......................... 105

L

Leather Waste 1

M

Mechanical Properties..................... 1
Methyl Blue Dye........................... 93

Multiple Time Scale Method 105

N

Nanoparticles 13
Nanostructures 61
Non-Local Theory 83
Nonlinear Vibration 105

P

Parametric Excitation................... 105
Physical Aging............................... 73
Picric Acid 47
Polyaniline Polymer...................... 93
Polylactic Acid.............................. 73
Polymers 13
Polymethyl Methacrylate Polymer 93
Porosity 83

R

Resin 1

S

Scale Effect................................... 83
SEM 47
Sensoristics 61
Sonication 47

T

Thermal Load............................... 105
Thermal Properties........................ 13
Thermally Stimulated Depolarization
    Current....................................... 73

W

Waste Jute Fabrics 1

Author Index

A
Adda, H.M. 83
Ahmed, S. 1
Al-Dabagh, S.Y. 93
Al-Hassani, E.S. 13
Ali, B. 83
Ali, M.F. 1

B
Benrekaa, N. 73
Bouamer, A. 73

C
Chowdhury, A.M.S. 1

D
Di Sia, P. 61

G
Ghadiri, M. 105
Gupta, A.K. 35
Gupta, S. 35
Gupta, S.K. 35

H
Hassan, S.M. 93
Honey, S. 47
Hossain, M.S. 1

J
Janmohammadi, M. 105

M
Marwa, M.A. 93
Mohammed, A.A. 13

N
Nasir, H. 47

O
Oleiwi, J.K. 13

P
Pandey, A. 35

S
Sadatsakkak, S. 105
Saleh, W.R. 93

Shah, S.S.A. 47
Shayestenia, F. 105
Slimane, M. 83

Y
Yadav, R.K. 35
Younes, A. 73